# FIRE!
## PREVENTION
## PROTECTION
## ESCAPE

# FIRE!
## PREVENTION
## PROTECTION
## ESCAPE

by Bill Cantor

**World Almanac Publications**
**New York, New York**

To Jeanne, Dave and Bill

The character of a man can be easily judged by the way
he treats those from whom he may not benefit.

Interior design: Clare Ultimo
Cover design: Donald DeMaio
Interior art: James Meddick

Copyright © 1985 by Bill Cantor.

First published in 1985.
All rights reserved. No part of this book may be reproduced in any form or by any means without permission in writing from the publisher.

Distributed in the United States by Ballantine Books, a division of Random House, Inc., and in Canada by Random House of Canada, Ltd.

Library of Congress catalog card number: 84-51048
Newspaper Enterprise Association ISBN 0-911818-69-3
Ballantine Books ISBN 0-345-32190-1

Printed in the United States of America

**World Almanac Publications**
**Newspaper Enterprise Association**
**A division of United Media Enterprises**
**A Scripps-Howard Company**
**200 Park Avenue**
**New York, NY 10166**

# Dedication

This book is dedicated to the memory of Joseph Coopersmith whom I was honored to have had as a close friend and professional colleague. He was a brilliant man, a social scientist, born twenty five years ahead of his time. His heart and mind were constantly focused upon improving the lives and dignity of all. In addition to a great many other contributions, he spent an important part of his life unselfishly giving to the development of home fire safety and training programs.

I dedicate this book also to the memory of Fire Chief Jay W. Stevens, with whom I spent countless hours learning about the problems and sensitivities of fire safety education. Chief Stevens was known to the members of his profession, from coast to coast, and he was made honorary Fire Chief of 256 cities across the country. For 33 years he was Executive Secretary of the International Association of Fire Chiefs and later became the Director of Fire Prevention for this organization. His long service included: Fire Prevention Chief for the National Board of Fire Underwriters (38 years); Secretary/Treasurer of Pacific Coast Association of Fire Chiefs (30 years); California's first fire marshal (17 years). Chief Stevens devoted his life to fire safety and fire education. His contribution to the fire service is a monument to his memory. He will not be forgotten.

# Table of Contents

|  | Dedication | v |
|---|---|---|
|  | Foreword | viii |
|  | Preface | x |
|  | Introduction | xii |
| CHAPTER 1 | The Nature of Fire | 3 |
| CHAPTER 2 | "Hot Spots" at Home | 8 |
| CHAPTER 3 | Fire Prevention in the Home | 12 |
| CHAPTER 4 | Animals and Fires | 15 |
| CHAPTER 5 | The Home Escape Plan | 17 |
| CHAPTER 6 | Escaping a Fire | 23 |
| CHAPTER 7 | The Young and the Old | 28 |
| CHAPTER 8 | Hotels, Motels, and High Rise Buildings | 34 |
| CHAPTER 9 | Seasonal Fires | 39 |
| CHAPTER 10 | Fire Safety and Recreation | 50 |
| CHAPTER 11 | Fighting a Home Fire | 57 |
| CHAPTER 12 | Burns and Other Injuries | 66 |
| CHAPTER 13 | After the Fire | 71 |
|  | Index | 77 |

# Foreword

Why the need for a book about fire safety?

Some years ago, I worked in the fire alarm and security systems industry. Concurrently, I served as Acting Director of the Institute for Home Fire Safety (a privately supported national organization) and worked with Chief Jay W. Stevens, its Director of Home Fire Safety Education. I lived and breathed fire safety in the home 365 days a year—making myself and my staff available to inform the public about the everpresent danger of fire and how to prevent, protect, and escape that danger.

Year after year the vast number of fires continued, along with fatalities, all tragic and so many needless. The advent of smoke detectors brought a decrease in deaths, but injuries and the number of fires continued to increase. During this early career and even now, I see an increase in the amount of flammable items used in clothing and construction and see no consolidated effort by any individual, group, or organization to do much of anything about flammability. In addition, there is little done about the toxicity of burning materials. The poisonous fumes from them cause the majority of fire fatalities. Fire safety and prevention programs lack the required funding at the national, state, and municipal levels and there are many who justifiably fear this ever present threat from fire.

Our municipal and volunteer fire departments and their valiant firefighters are our second line of defense against fires—after they have broken out. You and I are the first line of defense against the occurrence of fire. Hence the reason for this book. I hope that the information that is presented here will raise your consciousness about the ongoing threat of fire, what you can do to help prevent it, how you can protect your family and yourself against it and what to do in the event a fire breaks out at home, at work, or while you are traveling.

## Foreword

I don't claim to have all of the answers to every fire situation that might occur, but I believe that this book will serve a valid purpose— and save lives.

I want to thank Edward Hymoff, who has taken a mass of material and my thoughts about fire safety and helped me present these safety tips to you.

I also want to thank Jeanne, my wife and partner, for her love, devotion, patience and support and for putting up with my frustration over the rate of tragic deaths and injuries caused by fire, which has led me to write this book.

# Preface

Fire has plagued us from the beginning of time. Unfortunately its terror has been part of our human heritage. The Chronology of U.S. Fires is a short review to spotlight America's history of fires that have left their trail of tragedy.

### Chronology of U.S. Fires, 1608-1980

| | |
|---|---|
| 1608 | The settlement at Jamestown (Virginia) is completely destroyed by fire |
| 1788 | Eighty percent of New Orleans is destroyed by fire. |
| 1805 | Fire devastates Detroit. |
| 1871 | Chicago goes up in flames because of Mrs. O'Leary's cow—or so the story is told. |
| 1886 | Portland, Maine, loses half its buildings to a massive fire. |
| 1911 | In New York City, the fire at the Triangle Shirtwaist Company is the nation's first serious workplace disaster. More than 140 workers lose their lives. |
| 1918 | Minnesota loses 26 towns to a fire. |
| 1942 | The Cocoanut Grove nightclub fire in Boston claims 492 lives. |
| 1947 | The Texas City explosion and fire kills 500 people and levels the city. |
| 1980 | The MGM Hotel blaze in Las Vegas kills 84 people. In White Plains, New York, a fire at the Stauffer Inn kills 26 businessmen. |

Stories by the thousands tell of injury, disfigurement, and death caused by fires. Perhaps less spectacular, but nonetheless tragic, are the numerous fire stories aired on television and reported in newspapers throughout the United States every day.

What does not receive equal media coverage to the MGM fire or the Stauffer Inn blaze is this statistic: in the period from January 1979 through January 1980, there were 11,550 hotel or motel fires reported

# Preface

to local fire services. No one knows how many more went unreported, although logic dictates that the number may be equally shocking. No hotel or motel wants negative publicity, and so if there is a small fire, they will avoid calling the fire service whenever possible.

The horror of fire faces us each and ever day. What can we do to reduce the incidence of fires? And why haven't we done it? On the second question, we can cite several contributory factors:

1. **Apathy.** My home is built well. I've never had a fire, so I must be doing something right. I've got fire insurance, so fire can't hurt me! Sound familiar? Besides fire is too horrifying to contemplate, so it is easier to dismiss it as happening to someone else, somewhere else—in a ghetto, for example. (Fact: there are fewer fires in ghettos than in more affluent areas.)

2. **Education.** Not enough information reaches the public. Fire services do not have the money, time, staff, and expertise to educate people about fires. They have their hands more than full just doing what they do so well—extinguishing fires and rescuing those trapped in burning structures

3. **Visibility.** Although fire is the third largest cause of accidental injury and death in America, no politician, community leader, famous entertainer, or athlete has chosen to publicly commit his or her time, effort, reputation, or other resources to a total fire safety program.

The National Fire Data Center estimates that 32.6 million fires occur a year in the United States. That averages out to about 87,000 fires a day, or more than 3,600 fires an hour. It means that there's at least one fire every second!

More than 3000 fires are reported each day and no one knows how many fires go *unreported*. Nevertheless, most Americans do not know how to save themselves in the event of fire.

It is my sincere hope that the information contained in this book, written as it is in a simple, readable style, will help you prevent fires, protect yourself and your family from unsafe conditions that may start a fire, and show you how to help escape harm if a fire strikes.

I extend my deepest appreciation to Jane Flatt, publisher of *The World Almanac,* for supporting the publication of this book

# Introduction

Some astounding facts and figures appeared several years ago in a document prepared by the National Fire Data Center for the Federal Emergency Management Agency/U.S. Fire Administration.

- The United States and Canada continue to have the highest fire death rates in the world.
- No disease kills and injures more children than fire.
- The total number of reported and unreported fires in the United States over a one-year period is estimated at 32.6 million.
- The number of injuries caused by fire in the United States may be as high as 2 million a year.
- In the United States, twenty times more deaths are caused by fire each year than by floods, hurricanes, tornadoes, and earthquakes combined.
- One out of every eight accidental deaths in America is caused by fire.
- There are over 300 different ways a fire can start in your home.
- Seventy-five percent of all Americans who die in fires die in home fires.
- Among accidents that cause death and injury, fire is the third largest killer in the United States, according to the National Safety Council.

No one is immune from fire! Every year, fire statistics tell a sad story of deaths and injuries and more than $2 billion in property losses, including the destruction of over 700,000 homes. The number of fires and deaths from fires increases each year.

A fire can occur anywhere and at any time. Newspapers, radios, and television constantly report fires. News reports of fires are usually stories about fatalities. And although most fire deaths occur in the home, many fatalities have occurred in workplaces and hotels.

# Introduction

Fires and the threat of a fire are always in season, but some "seasonal" fires require special vigilance. For example, in the summer months we must be especially concerned with the proper use of barbecues and campfires. Autumn is the time to take care with the burning of leaves and brush. Improperly used candles and jack-o-lanterns and the wearing of flimsy nonfire-retardant costumes cause injury and death on Halloween. The Christmas season has its own history of tragedies caused by the careless use of holiday decorations.

Fire can happen to anybody—rich, poor, young, and old. Fire strikes affluent suburbs as well as poorer inner cities. In recent years a number of celebrities have had encounters with fire. The list includes former President Richard Nixon, former First Lady Rosalyn Carter, Vice President George Bush, comedienne Carol Channing, singer John Davidson, Senate Majority Leader Howard Baker, heart surgeon Dr. Michael DeBakey, baseball star Dave Winfield, and actor/dancer Gene Kelly.

Each one of these well-known individuals escaped a life-threatening fire. President Nixon and Mrs. Carter were among the world's most heavily guarded and protected people when the fires in which they were involved occurred. Perhaps there is a message for us in fires to which prominent people have been exposed. We may think that fires always happen to "somebody else." But if such danger can threaten a President and a First Lady, anyone can become a fire victim.

Ask yourself these questions:

- Do I know what safety measures to take to help prevent a fire?
- Can I save myself and my family if a fire breaks out?
- Does my family know how to save themselves should a fire occur?
- Do I know what to do if a fire occurs at work or at a hotel where I am staying?

Answers to these vital questions rest with our knowledge of fire prevention, protection, and escape.

A number of organizations are working around the clock to protect you and your family from injury or death by fire. They include your local fire department, public safety organizations, and commercial firms. Your fire department, for example, may be affiliated with the International Association of Fire Chiefs and the International Association of Fire Fighters. Another organization is the National Fire Protection Association (NFPA), since 1896 the world's principal authority for promoting the study and application of fire science and improving methods of fire prevention and protection.

# FIRE!
## PREVENTION
## PROTECTION
## ESCAPE

# Chapter 1

# The Nature of Fire

Few of us understand fire. We know enough not to strike a match when a sign warns us that it's dangerous to do so or when "Gasoline Is Stored Here." But there are other fire dangers too.

How many of us are aware that there are more than 300 ways that a fire can start in the home? All it takes are three ingredients for a fire to start:

$$Fuel + Heat + Oxygen = Fire!$$

When these three ingredients are joined and react together, fire occurs! It's that simple. It's that dangerous. Take away any one of the three ingredients and a fire cannot occur. For example, heat and oxygen add up only to hot air. More simply put:

$$Fuel + Oxygen = 0$$

We are surrounded by dangers that can fuel a fire. They are in our homes, at our workplaces, and wherever we spend our leisure time. The average home, for example, has at least one to two *tons* of "fuel" in the living room—upholstered furniture, window drapes, rugs, bookshelves, magazines, newspapers, and lamps.

There is more "fuel" in the average bedroom—in such things as bedding, dressers, linens, and window drapes. The attic, often used for long-term storage, contains discarded items that in the heat of summer, can dry out to become the "fuel" that awaits the two other ingredients that cause ignition—heat and oxygen. The cellar and garage, especially if the garage is joined to the home, often become storage rooms for items that can fuel a fire.

The uncontestable fact is that there is a constant threat of fire around us. A fire can begin in so many ways that it's impossible to predict how,

### Fire! Prevention, Protection, Escape

when, or where the combination of air, heat, and fuel may ignite and fan a fire that will destroy property and injure and kill people.

The poisonous, superheated air that can be created by a fire has been measured at temperatures of 800 degrees to 2000 degrees. That's hot when you realize you can broil a chicken at 500 degrees to 600 degrees.

In a fire test of burning buildings conducted in Canada some years ago, instruments from within the burning buildings measured the movement of the poisonous, superheated air. It was clocked at speeds of 90 feet per second, or 60 miles per hour. These tests demonstrate that a fire leaves little time in which to react and that elements other than flames can kill. Many fires would never start if safe habits replaced dangerous ones. Let's take a look at some of the dangerous habits and see what can be done about them.

## Correcting Dangerous Habits

### Matches

The misuse of matches has caused fires in the home, office, workplace, and out-of-doors; in vehicles and in boats. Matches that were carelessly tossed aside, left where children could find them, or thoughtlessly forgotten have been a major source of fire.

These simple rules for using and storing matches should always be observed:

- Keep matches out of the reach of children.
- Don't dispose of a match until it is extinguished. Absent-minded adults strike and use matches and then toss them away immediately after use. This dangerous habit is a serious fire hazard.
- Strike-anywhere matches, especially if poorly made, often appear to be extinguished when discarded. This poor-quality match could ignite by friction; when carried loosely in pockets, such matches have been known to burst into flame from the friction created by the normal movement of clothing.
- Always keep matches in metal containers. After use, break them in half before disposal. Attention to this small detail can ensure that a match is extinguished.

### Household Cleaners

Careless household habits can cause fires. For example, improperly stored waxing and polishing cloths and mops can contribute to a fire that starts by spontaneous combustion—the effect of heat generated by oxygen and chemically treated materials causes them to burst into flames. The oils in paint oxidize and can ignite spontaneously. This

# The Nature of Fire

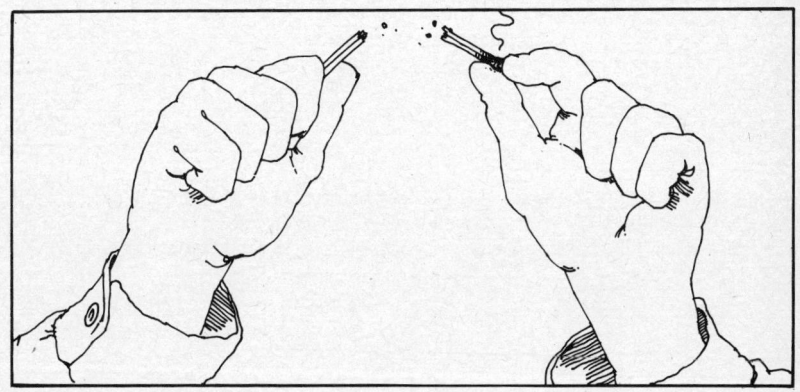

chemical combination develops heat that increases to the point where chemically saturated cloth fibers can ignite in poorly ventilated areas. Once this highly flammable mass is burning, everything in the confined area is consumed.

Cleaning rags, polishing cloths, and mops should be stored in tightly capped metal containers so that the oxygen cannot provide the key element in causing a spontaneous combustion fire.

Floor waxes, furniture polishing liquids, general household cleaning fluids, paints, varnishes, and paint removers are prime candidates for causing fires if they are improperly stored. These items should never be stored in glass jars, which can break. The safest procedure is to store them outside in their original containers and away from any source of extreme heat.

## Storage Areas

Storing sentimental treasures in attics, garages, cellars, or in that extra room quickly becomes the catchall family "warehouse", cluttered with old newspapers, books, cartons, odd pieces of furniture, and cast-off clothing. These items represent a fire hazard. Spontaneous combustion can easily occur among tinder-dry items stored in enclosed places for long periods.

Bare electric light bulbs hanging in attics and cellars and other storage areas in the home are also a danger. A hot light bulb gives off nine times more heat energy than light energy; if a hot bulb touches old clothing hanging from a nail or on a rack, it can cause a fire. Often, matches and candles are taken into such storage areas to add light to dark, cluttered corners. A candle that tips or falls can start a fire if the

**Fire! Prevention, Protection, Escape**

flame comes in contact with dry paper or other flammable items.

Anything stored in the house for a long period is easy prey to spontaneous combustion and should be stored in a proper container. Actually, the greatest safety measure for collectibles is to cart them off to your favorite charity or local antique shop. Uncluttered areas in the home provide additional fire safety for residents.

## ✔ Fire Safety Checklist for Eliminating Dangerous Habits

☐ Are the matches in your home stored in a small metal container and out of the reach of children?
☐ Do you make sure matches are extinguished before discarding?
☐ Are you careful not to carry loose matches in your pocket?
☐ Are all the cleaning rags and polishing cloths you have used stored in a closed metal container and away from extreme heat?
☐ Have you checked all cleaning fluids, waxes, and sprays for leaking containers?
☐ Have you stored all paints and varnishes in their original, tightly closed metal containers, and did you take similar precautions for paint removers? Are rags used in painting stored in closed metal containers? Are all chemically based materials stored in a well-ventilated area outside the home?
☐ Have you checked the clutter in your attic or other storage areas to determine what can be discarded: newspapers, books, magazines, furniture, clothing, and the like?
☐ Is the lighting in your attic or cellar a dangerous bare bulb? If so, does it hang away from stored items?

## Chapter 2

# "Hot Spots" at Home

## The Kitchen

Virtually all activity in the home revolves around one important area: the kitchen. In some homes the kitchen is in use from early morning until late at night when the last person to go to bed switches off the kitchen light.

Some thought must be given to this center of activity. The kichen is an area waiting for an accident to happen. The average kitchen can be a booby trap.

In fact, the kitchen should be looked upon as one of the key "hot spots" in any house or apartment. The kitchen stove frequently is the cause of a home fire. An example of an incident waiting for a fire is the simple process of cooking french fried potatoes. In one scenario, a woman sets a pot of cooking oil on the stove to heat. Suddenly, a child in a nearby bedroom awakens from a troubled nap and begins to cry. The woman stops what she is doing and rushes to her child's side. She has forgotten about the pot of oil cooking on the stove.

The cooking oil overheats. It bubbles and spurts, or it boils over and touches the hot elements on the stove and bursts into flames, splattering the entire kitchen with fiery oil. The fire and smoke attract the woman's attention. She grabs her child and in near panic dashes out of the home.

That is one scenario. And the mother and child were fortunate. This fire occurred in the three or four minutes that the stove was unattended by a distracted mother. Ordinarily she would have given her cooking undivided attention. If she had been minding her cooking, she could have controlled the fire by turning off the burner and placing a cover over the pot. This would have cut off the air supply, smothering the fire.

Among the many kitchen accidents that happen are those associated with cooking oil gone wild. (Cooking oil that splatters on the skin can

## "Hot Spots" at Home

also cause painful burns.) When this happens, immediately turn off the burner. A common household item like baking soda—not baking powder—spread on the fire will quickly extinguish the flames. Fires in ovens can be extinguished by turning off the oven and keeping the oven door closed to prevent air from fueling the flames.

Here are some safety rules that are among the best defenses against a fire in the kitchen.

- Replace long, flimsy drapes with short curtains that cannot blow against hot stove tops and ignite.
- Always work in the kitchen with sleeves rolled up. Loose sleeves can touch a hot area on the stove and catch fire. Wear a protective apron.
- Keep chairs and stools away from the stove and work counter to prevent toddlers from climbing onto the stove. Reminder: store matches in metal containers out of the reach of children.

Fire! Prevention, Protection, Escape

- Watch your electric appliances. They can cause fires under any number of circumstances, especially if they are connected to lengthy extension cords. Toasters and other kitchen appliances should not be used near curtains, wall-mounted towel racks, or low cabinets.
- Pay attention to cooking procedures when using the stove.
- Keep baking *soda* at hand at all times. Note that baking *powder* can be a kitchen hazard. Baking powder can flare up if used to put out a fire. *Do not confuse the two substances,* especially in a fire emergency.

## The Electrical System

Another "hot spot" in the home is the electrical system. According to the National Fire Prevention Association's *Fire Protection Handbook,* four out of five homes are underwired. Underwired homes lack adequate circuits to handle the large number of modern electric appliances. Electrical equipment that can ignite floor coverings or upholstered furniture is one of the prime causes of fire in the United States, according to the NFPA. Many home fires caused by electrical wiring have been traced back to do-it-yourself wiring and electrical work, where the workers were not as knowledgeable as they should have been. When they did not use the correct-size wire, they overloaded the circuits and created a fire hazard.

One electrically caused fire occurred when a light fixture was improperly wired in a garage attached to a $425,000 home. This fire resulted in heavy property damage.

Another electrical fire broke out in a home when a television set connected to a maze of nearby wires suddenly shorted. The television set began crackling and popping, emitting the acrid black smoke common to burning circuits and plastic. The inhabitants fled to safety.

In yet another electrical fire, a hot spark arcing from the base of the electric meter started a fire under the new aluminum siding of a house. The fire continued, burning the dry shingles under the aluminum siding. A neighbor saw the smoke and called the local fire department. But the resident of the burning home, who was sixty-five years old, was overcome by smoke and died of a heart attack despite rescue by the volunteer firefighters.

Electrical fires are unpredictable. They are high on that list of 300 different ways that an accidental fire can also provide peace of mind, for they include correcting faulty wiring.

- Are your circuits overloaded with "octopus" outlets that permit many electric cords to be plugged into a single outlet? If so, disconnect.

"Hot Spots" at Home

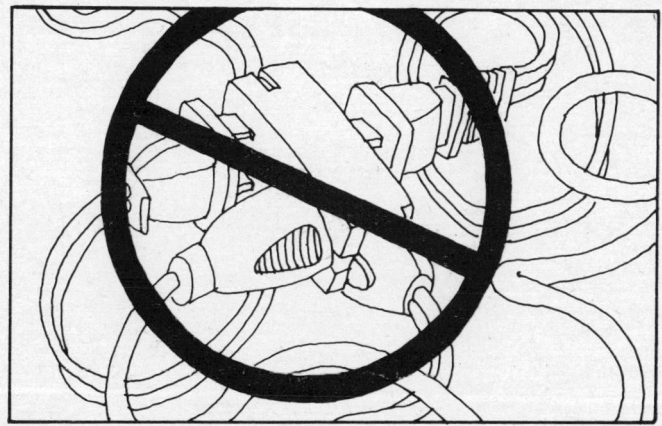

- Are there frayed electric cords on any of your appliances? Do *not* use appliances with frayed cords.

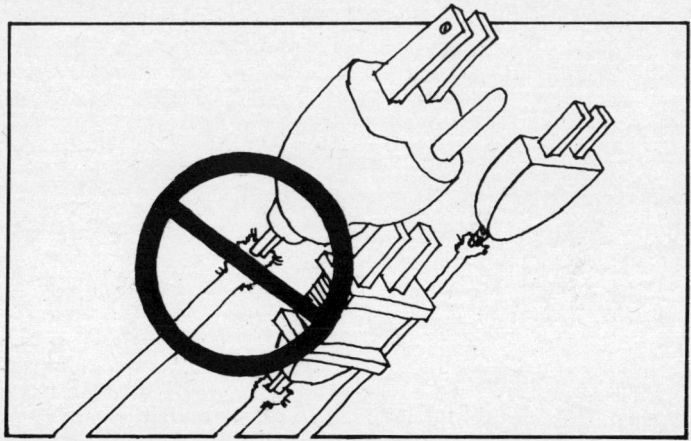

- Are your hot electric appliances kept away from items that can easily ignite, such as curtains, towels, paper napkins, and newspapers or magazines?
- Do you regularly clean all electric motor-driven appliances, especially the motor air vents used for cooling the motor, which can easily clog with combustible dust or lint?
- Is all the wiring in your home certified by a licensed electrician?

Chapter 3

# Fire Prevention in the Home

Fire prevention must begin at home. Then we must extend our knowledge of fire prevention and safety outside the home, for the threat of fire constantly follows in our footsteps.

Of the four kinds of fires classified by fire prevention experts, three occur in homes. All four classifications lurk in the workplace, waiting for that "accident" or careless "mistake" to happen.

Class A fires are caused by the ignition of ordinary combustible materials such as wood, paper, cloth, rubber, and many kinds of plastics.

Class B fires involve flammable and combustible liquids, gases or fumes, and greases.

Class C fires begin when "live" electrical equipment malfunctions.

Class D fires involve combustible metals such as magnesium, sodium, titanium, and potassium, which are usually found in the factory workplace and are not normally found in the home.

Let's take another look at the home "hot spots" we discussed in the last chapter.

## The Exhaust System

Another kitchen hazard is the exhaust system over the stove. Many modern kitchens have electric-powered exhaust fans that carry grease fumes and smoke from cooking out of the home. If this exhaust system is so prevalent in homes, why is it a potential hazard?

According to analyses of kitchen fires, the hazard is the grease fumes that settle on the metal fixtures in the exhaust system. Over a period of time these grease fumes build up a thick and hard residue of a sticky, oil-like substance that softens when heated and becomes a fire hazard.

Droplets of grease that strike flame or fall on the heating element of an electric stove tend to "explode" in miniature fireballs. These fireballs can ignite dripping grease. A grease fire can reach the exhaust fan grille

## Fire Prevention in the Home

with its accumulation of soft, flammable grease. This accumulation can burst into flames. If the exhaust fan is switched on, fire can be sucked into the ventilation ducts, which spread the grease-fueled flames farther.

A grease fire in a kitchen exhaust system is difficult to extinguish by conventional means. If possible, the exhaust system should be switched off, the house evacuated, and the fire department summoned—in that order.

There is a simple way to help prevent such a fire. The exhaust system should be cleaned at regular intervals. Cleaning the grille, fan blades, and ducts two or three times a year can provide that extra measure of safety in the home.

## Vapor Fuse Fires

Another common fire hazard in the home is a "vapor fuse," an invisible trail inadvertently created by fumes from flammable and volatile substances such as gasoline, kerosene, and naphtha. These substances are used in house painting, in liquids or pastes to clean floors or furniture, and in other materials used in the home. Volatile and highly flammable liquids rapidly evaporate at room temperature. The fumes or vapors rise from the open containers and can be touched off by a spark or flame. The trail of fire can lead right back to the source. Flammables in containers often have been known to explode when the vapor trail fuse is ignited.

One such vapor fuse fire occurred when the owner of an apartment building began using a volatile liquid to prepare interior wood surfaces for painting. Not recognizing the danger, he worked without opening windows for proper ventilation. When he switched on an electric light, the electric spark met the volatile liquid's vapors in a blinding explosion. A fireball created by the ignited fumes touched off wood surfaces that

### Fire! Prevention, Protection, Escape

had been treated with the liquid. All wood surfaces caught fire. Not even the best efforts of firefighters, who arrived at the scene quickly, could save the structure.

Preventable fires caused by vapor fuse occur out of sheer ignorance or through carelessness. Volatile and highly flammable liquids such as paint remover, benzene, turpentine, and gasoline evaporate at room temperature. The only difference between the open container of liquid that gives off dangerous volatile fumes and a sputtering fuse attached to a stick of dynamite is the speed at which the vapor fuse can ignite. Experiments have shown that a vapor fuse can explode in a fraction of a second.

Volatile and flammable liquids should always be used in well-ventilated areas and should be stored in metal containers. The safe way to store such liquids is in safety cans bearing the safety seal of a national testing laboratory. These liquids should be stored outside the home.

## Commonly Used Flammables

Some of the flammables in our home are everyday items that we take for granted. The average bathroom medicine cabinet can contain a variety of potential fire hazards. Many medical prescriptions and off-the-shelf medications purchased in drugstores contain volatile and flammable liquids.

Cosmetics and other beauty aids are potential fire hazards. Hair sprays, fingernail polish and polish remover, and cosmetics that contain acetone, lacquer, or an alcohol base are all ingredients that can contribute to a fire.

Medicines and cosmetics should be tightly capped and stored in metal medicine cabinets.

---

### ✓ Fire Prevention Checklist for Hazards in the Home

- ☐ What is a Class A fire? A Class B fire? A Class C fire?
- ☐ When was your kitchen ventilation system last checked for accumulation of grease?
- ☐ Is your exhaust system due for a cleaning? If not, mark your calendar for when the next cleaning should be scheduled.
- ☐ Do you know how to store your cosmetics?
- ☐ Do you open windows and doors to prevent a vapor fuse when using volatile and highly flammable liquids?

## Chapter 4

# Animals and Fires

Birds do it. Squirrels do it. And cats and dogs do it. Do what? Accidentally start fires, of course!

The following stories are true. They have been taken from the records of various fire departments.

A pet cat, kept in a drugstore to prevent rodents from settling in the building, started a fire when it began sharpening its claws on a box of wooden matches. The cat's claws cut through the flimsy wood and scraped against the match heads, causing them to ignite. The box of matches burst into flames and set fire to other materials. Damage was estimated at $35,000.

Two frolicking Siamese cats in the living room of one home knocked over a lighted floor lamp, which lost its lampshade. The hot, bare, 100-watt bulb fell against the window drapes. The drapes caught fire. Flames spread to upholstered furniture, causing thick, black smoke laced with highly toxic gases. The family—both parents and their two children—was in a second-floor room watching television. Unable to escape, the four members of the family and their pet cats succumbed to the smoke and toxic gases created by the burning upholstered furniture.

In a recorded yet unbelievable fire, a sparrow picked up a discarded, smoldering cigarette in its beak and flew to a nearby wooden building, which housed the public library. The sparrow settled itself on a canvas awning on which it had made a nest. It lined the nest with the lighted cigarette. The dry twigs caught fire and in turn set the awning ablaze. This was followed by the wooden structure's also catching fire. The town's public library burned to the ground.

In still another case of an animal causing a fire, a squirrel picked up a smoking cigar butt, carried it up a tree, and dropped it. The cigar fell on dry autumn leaves at the base of the tree. A nasty backyard fire was started, and two homes were destroyed.

Another documented fire occurred when a dog knocked over an open

### Fire! Prevention, Protection, Escape

can of kerosene in a barn that was being converted into a home. The flammable liquid splashed against a hot kerosene heater, and the barn went up in flames with a loud *whoosh*. The plans for converting the barn into a residence also went up in smoke. Luckily, there were no injuries.

One recorded fire began after a mouse crawled inside the wall of a wooden structure faced with brick. The curious rodent chewed some wires in an electrical box, which caused a short circuit—and a fire. It also caused the death of the mouse.

In another fire involving a bird, a house was destroyed after a bird built a nest between the structure and the outside chimney. The chimney overheated and set fire to both nest and home.

Also recorded are unusual fires caused by a freak accident, or, as some home fire insurance policies describe them, by an "act of God." In one such case, a golfer was hit on the back pocket of his trousers by a golf ball. The ball actually slammed into some loose "strike anywhere" matches that the golfer was carrying in the pocket. A fire was started by the sudden, intense friction of the matches, which also created a "hot seat" for the golfer.

In other case studies of fires, two similar incidents occurred that defy belief. In one incident, a fire began after the winter sun's rays focused through an icicle hanging from the eaves of a snowbound house. The magnified sunlight speared through the window and onto a table in the dining room, setting fire to the tablecloth. The second incident occurred in a baby's carriage. The infant's empty glass milk bottle was lying in direct sunlight, which turned it into a magnifying glass. The concentrated sunlight on the baby's light blanket set the blanket afire. The baby was seriously burned.

Weird, yes. Unusual, yes. But fire can break out under the strangest of circumstances. Under control, fire is man's best friend; out of control, it is his worst enemy.

## Chapter 5

# The Home Escape Plan

A fire breaks out in a home at night. The smoke detector alarm goes off and wakes the family. A passerby telephones the local fire department. The family evacuates the house. Minutes later, the members of the family assemble outside in the street as the head of the house takes a head count. Everybody is accounted for. The whine of a siren can be heard even before the first fire truck arrives. Some firefighters pull out hoses to attach to a nearby fire hydrant, while others warily move toward the house, which is bathed in floodlights from the fire trucks.

Other fire engines arrive. Drivers direct their floodlights toward the house, from which smoke is billowing. Seen in the smoke is the orange glow of the fire, burning brighter with each passing minute. The firefighters attack, directing fingers of water at the burning structure. Others close in toward the building.

Meanwhile, neighbors clad in pajamas covered by bathrobes look out their windows or leave their homes and move toward the helpless family burned out of its home. The neighbors offer help and sympathy as they ask how the fire started. The firefighters are too busy to pay attention to the victims, who have safely escaped the fire. The fire chief already has ascertained that all members of the family are accounted for.

This family was fortunate. They had prepared for and followed a planned fire escape. Later that night, in another part of town, another family would not be so fortunate. Three people would die because the family had no escape plan.

Three quarters of the people who die from fires in their home each year die because they had not developed an escape plan. After all, fires always happen to "somebody else."

When a fire occurs in the home, there's always the possibility that people caught in the fire will panic. There is just no way to predict how people will react during a fire, just as there is no way to predict, when, where, and how a fire will break out. This is why your fire department is on call 24 hours a day, 365 days a year.

17

Fire! Prevention, Protection, Escape

A carefully prepared home escape plan is not complicated. In any emergency situation there is little or no time to become involved in complicated procedures. For example, one elderly lady had three locks on her door for protection. One was a combination lock, and the other two required different keys. On the night the fire occurred in her home, she had to fumble in the darkness for two of the keys needed to unlock her front and rear doors. The fire had rendered the electrical system useless, as a fire often will, and she probably couldn't see the combination lock without a light. When firefighters broke in, they found her body at the front door. She had a key ring clutched in her hand. Apparently she had never developed or practiced an escape plan through either the front or the back doors. She paid with her life for this oversight.

Along with helping to prevent a fire from starting are two other important points: protection and escape. Prevention and protection go hand in hand. Being prepared to escape from a fire is also essential.

Some aspects of fire prevention were discussed in earlier chapters. Fire protection and fire escape also go hand in hand. There are many sophisticated home fire alarm systems on the market today. Fire protection requires, at the minimum, at least one smoke detector in the home. If only one detector is used, it should be strategically placed in front of the entrance to the bedrooms. All smoke detectors should be checked regularly.

## Your Home Escape Plan

Exactly what is a home fire escape plan? First, it should be a well-thought-out written document. Your life and the lives of your loved ones may depend on the plan that you have developed. The best way to start a home escape plan is to spend the necessary time to study the layout of your home.

On graph-lined paper, or any paper for that matter, draw the dimensions of each room as if you were looking at the layout of your home with the roof taken off. Note where the windows and doors of each room are located and how the furniture is placed. Show the corridors and stairs. Indicate any window from which you can easily reach a roof below.

Give some thought to what it would be like if a fire started in a bedroom, the living room or the kitchen. For example, imagine that there is no possibility of evacuating the house by using the inside stairs to the first floor or by the front or rear doors. In this case, the only way out is through a window. Think about this possibility. Ask the rest of the family to join in this exercise. Then draw dotted lines, dashes, or unbroken lines that depict the safest and quickest routes out of the house in the event of a fire. There should be a primary escape route and a backup, or alternative, escape route.

This document is one part of your escape plan. It's an important element because it's a map to safety.

The Home Escape Plan

## HOW TO MAKE YOUR FAMILY ESCAPE PLAN

1. Draw a floor plan that includes every room.
2. With your family, visit each room and plan an escape from that room.
3. Plan an alternative route in case the normal route is blocked by fire or dense smoke. Draw alternative routes in *different colors*.

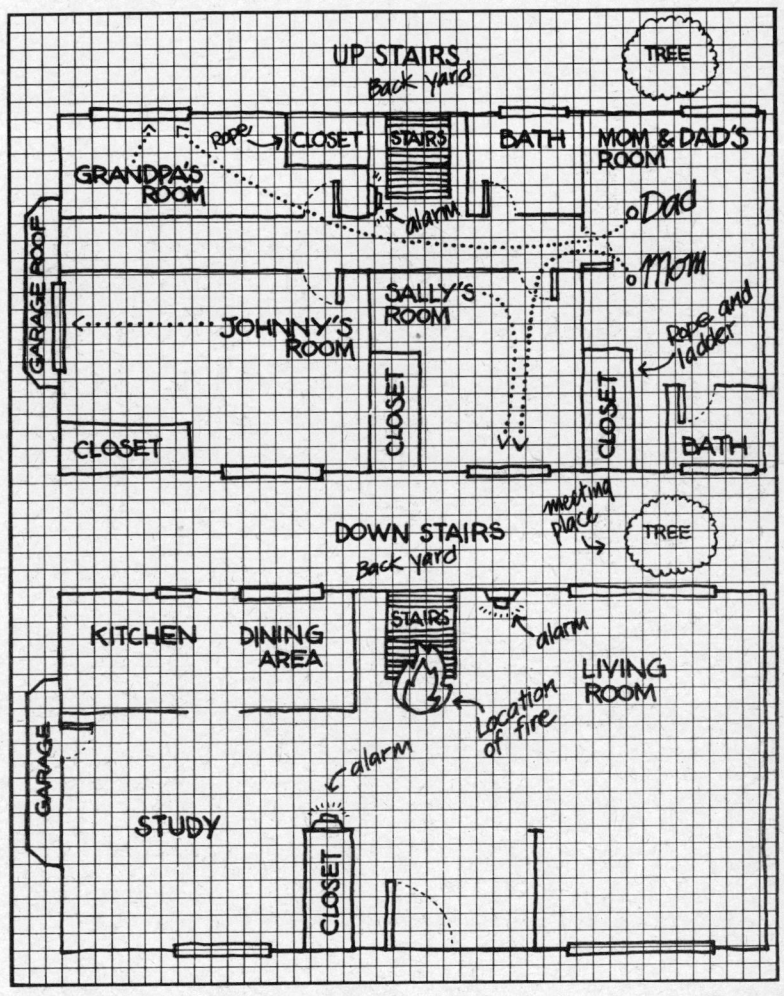

## Fire! Prevention, Protection, Escape

## ESCAPE PLAN

Fill in your complete route to safety outside the house from every room inside the house. Space is left beneath each room designation for you to fill in explanatory details as to room location and use.

| Room | Normal Route to Safety | Alternate Route to Safety | Emergency Aids or Measures (Rope, Ladder, Key, Etc.) |
|---|---|---|---|
| 1. Living Room | | | |
| 2. Study | | | |
| 3. Dining Room | | | |
| 4. Kitchen | | | |
| 5. Bathroom | | | |
| 6. Bathroom | | | |
| 7. Bedroom Mom & Dad's | Exit down stairway and out front or back door. | Dad goes for Grandpa, then exits out window on rope ladder. Mom goes for Sally. | |
| 8. Bedroom Johnny's | | Exit onto garage roof, then wait for help. | |
| 9. Bedroom Sally's | | Wait for Mom, then exit out window with help. Mom pillows. | |
| 10. Bedroom Grandpa's | | Wait for Dad to exit out window on rope ladder, then follow. | |

20

## Responsibility and Rehearsal

All occupants of the home should know what to do if there's a fire. Responsibility and assignments for the safety of all occupants should be determined when developing an escape plan. For example, teenage occupants could assist youngsters to evacuate the home. Parents could look to the safety of grandparents or other elderly occupants.

As soon as each person evacuates the house, he or she should assemble at a predesignated meeting place nearby—a tree in the backyard, for example, or next to the telephone pole on the front sidewalk. When possible, the meeting place should be in a well-lighted area.

An escape plan also should include two added activities:

1. One person should be responsible for calling the local fire and police departments. The telephone numbers of these departments should be made available to all family members, who should keep them handy.

2. The location of the nearest street fire alarm box should be known to all occupants of the home, and all should know how to activate the alarm.

There are no set rules for escaping a fire. Holding a fire rehearsal or drill several times a year is the smartest thing to do. An escape plan that is put off can be a costly mistake.

Every home is different. No two home escape plans are exactly the same. Even if some houses and apartments are laid out alike, each dwelling contains different furnishings placed in different ways. And as people differ in their preferences for certain kinds of furniture and where it should be located in the home, families also differ in age and number of family members.

No all-embracing home escape plan fits every home or apartment. I have tried to instruct you on how to go about developing *your* home escape plan. And after you have drawn your plan, I suggest that you show it to your local fire department. The department may be able to offer suggestions for the plan's improvement. Once you and your family have prepared an escape plan, you will have greater peace of mind in facing up to the danger of fire.

## ☑ Family Fire Escape Plan Checklist

☐ Have you developed your home escape plan?
☐ Do you have an alternative escape route?
☐ Who is responsible for which child if there is more than one child?
☐ Who is responsible for the elderly occupants of your home?
☐ Have you designated a meeting point outside in a well-lighted area so that someone can take a head count?
☐ Are the phone numbers for your local fire and police departments readily available and at each telephone?
☐ Does each occupant have a copy of the home fire escape plan?
☐ Have you held a home fire drill recently? Have you rehearsed your escape at night with the lights out, using only a flashlight?
☐ Have you also used the alternative escape plan?
☐ Have you corrected deficiencies in your escape plan?

## Chapter 6

# Escaping a Fire

No two fires are the same. Yet the fire safety rules are almost identical for fires that occur in single-family homes and apartment buildings.

Any escape plan should point out the "trouble escape." Can locked doors be opened easily? Do windows open easily? A "trouble escape" occurs when exits do not open in an emergency or when the planned fire escape route is blocked by fire-created or man-made barriers, the headway made by the smoke or flames, or both. A fire can gain headway before the full extent of the emergency can be established by those trying to escape.

## Doors and Stairways

An important safety barrier in the home is the closed door. At night, all bedroom doors should be closed. Surprisingly, a door is a barrier against the spread of flames or smoke. A tightly fitted door may not stop a fire, but it will slow the spread of flames and smoke, and can provide the extra time needed.

Closed doors should latch shut. Superheated air can create enough pressure to blow open an unlatched door. For example, a basement door at the top of stairs leading directly into a home or apartment building should always remain latched shut. Occupants should be trained to close and latch shut the basement door each time it's used. A fire in the basement often can be contained until an emergency escape is carried out.

Most modern homes are designed with an open stairway above the basement level that gracefully leads to the floors above. Although aesthetically pleasing, this is among the worst possible designs from a fire protection standpoint. In the event of fire, this design provides an unobstructed flue effect that can lead to a rapid upsurge of superheated air and smoke.

## Fire! Prevention, Protection, Escape

There are a few simple rules to follow if a building is on fire, no matter what kind of building it is. Do not open any door without first checking to see if the door is hot. Using the palm of the hand to feel heat, start at the top of the door and run your hand along the crack down toward the floor. If the door is very warm or hot, the cause could be superheated air and could indicate that there's fire burning on the other side of the door. An alternative escape route should be taken immediately.

If you're uncertain about the temperature of the door as you run your hand along the crack of the door, from the top down, carefully brace with one foot and press shoulder and knee against the door. Open the door about one inch. The bracing procedure will prevent the door from blowing open from superheated air, the one-inch opening will let you look out to see what's happening. If smoke quickly pours through the opening and pressure or warm air is felt from the other side of the door, quickly slam the door shut. Make certain that the latch catches. Pressure can blow open an unlatched door.

If no smoke or hot air is present, look through the slightly open door to see if it's safe to proceed.

If it's clear, then venture forth, closing the door so that it latches behind you. But make certain that the door doesn't lock in the event that a retreat is necessary, and reentry is needed.

**Escaping a Fire**

## Windows

Windows designated in an escape plan can cause problems. Studies have shown that conventional windows that slide up and down are prone to stick more often than doors fail to open. Alternative escape routes are often through windows. Periodically check all windows that are part of your escape plan. The best time to do this checking is during your fire escape rehearsal.

An ornery window will stick when you most need to open it easily. In a emergency, don't waste time struggling to open the window. Smash it! Use any nearby object—a baseball bat, for example, if the window is in a youngster's room. A man's shoe can be used. A metal lamp base or chair also can be used.

The safest way to break a window is to stand to one side with face turned away to minimize injury from flying bits of glass. Then clear away the dangerous shards of glass that might cut you on the way out.

To escape from a fire through a window, exit feet first with stomach pressing against the windowsill. If the window is no more than six feet above the ground, there is no problem for adults or tall teenagers. Young children can be lowered to safety. If the window is higher, an adult can hang full length and drop 10 or 12 feet without too much trouble. When falling to the ground in this manner, knees and body should be limp and a roll backward will help absorb additional shock. Windows higher than 15 or 20 feet require ladders. Falls from such heights can cause broken bones or worse.

**Fire! Prevention, Protection, Escape**

If caught in a high-rise fire, remember that the tallest fire department extension ladder reaches only to the ninth or tenth floor. Knowing this, you might wish to request a room below the ninth floor.

## Smoke

Encountering smoke during a preplanned escape calls for a retreat in the opposite direction. But if there's no route to safety except through the smoke, follow these basic precautions:

- Drop to hands and knees and crawl, keeping head down. This will help you to breathe the clearest air. Smoke and heat accumulate at the ceiling and bank downward.
- If water is available, hold a wet cloth or towel over mouth and nose. Although the wet cloth is not a good filter against toxic gases, it can help screen some irritants from the smoke and make breathing more comfortable. Take shallow breaths to avoid deeply inhaling toxic gases.
- Navigate through the smoke-filled area with eyes tightly closed, if possible, to avoid irritation and severe discomfort. Crawl slowly, using your sense of touch; occasionally open your eyes to determine correct line of travel; proceed until out of the smoke-filled danger area.
- If a smoke-filled stairway must be navigated, take the same precautions. Under the worst of smoke and heat conditions, crawl down backward.

## ✔ Escaping a Fire Checklist

- [ ] Have "trouble escape" routes been identified, and are they accessible?
- [ ] Have all windows and doors been checked to see that they open easily?
- [ ] Are all doors closed at night, and do they latch shut?
- [ ] Are there working flashlights in every bedroom? Are there rope ladders, keys, or other items available for escape where necessary?
- [ ] Who is responsible for calling the fire department?

―――――――――――― Chapter 7 ――――――――――――

# The Young and the Old

Fire discriminates against no one. It alarms the healthy and the sick, the old and the young, the rich and famous and the poor and unknown. But some age groups are at special risk.

## Children and Fires

Left unattended, children often cause fires. Each year, tragic fires started by children take their toll. The very young are curious, and matches or cigarette lighters left around the house are often pocketed by children who want to experiment with these new "toys" that they have seen used by parents or older relatives.

Children can be taught fire prevention by parents who have developed their own good safety habits. A child learns by imitating parents and older brothers and sisters. Parents should explain to young children how kitchen fires can be avoided. How, in the family workshop, certain procedures are followed to avoid starting a fire. These particular practices should be explained thoroughly and patiently.

Instructing children about the dangers of fire is a never-ending role for parents and older family members. Young children also should participate in the family fire escape rehearsals with their elder sisters and brothers. The reasons why these escape drills are held should be explained in detail in a calm and simple way that the child can understand.

Studies have shown that youngsters who have not participated in fire escape drills inevitably panic and often hide when a fire breaks out. Investigations of tragic fires that have been started by children show that a child's strongest instinct is to run and hide in a closet, under a bed, or in some other enclosed space. A child views tight confinement as a safe haven.

## The Young and the Old

In fires that have claimed children as victims, many of the youngsters were found in closets or out-of-the-way corners of the gutted home. More often than not, they had died of smoke inhalation.

A child who is too young to perform the escape procedure alone should be instructed to remain in one place when a fire alarm is sounded, until an adult or older teenager appears on the scene to give assistance. Once the youngster is able to use the normal escape route, then a simple rule should be established to leave the house and meet other members of the family at the meeting place. If the normal escape route is blocked, and the child cannot join an adult, the youngster should go to a window where he can be seen, call for help, and wait to be rescued.

By the time children are eight years old, the family's regular fire drills should have prepared youngsters to escape under most emergency conditions.

### Briefing the Sitter

Despite escape training and fire rehearsals, children should never be left alone at home—not even for a "few" minutes. Competent child sitters can usually be found if arrangements are made in advance.

Unfortunately, untrained sitters are often engaged to watch homes and children. Tragic accidents have occurred. One teenaged sitter's attention was directed to a television program instead of her four-year-old charge. The youngster found a book of matches and struck one as

### Fire! Prevention, Protection, Escape

she had seen her mother do. The book of matches ignited, and the little girl's dress caught fire.

The sitter panicked. Paralyzed at first by the sight of the child's burning clothing and her anguished screams, the sitter spent precious lifesaving minutes trying to fill a pan of water in the kitchen to douse on the little girl's clothing.

In the hospital the next day, the fire victim, who suffered painful and serious burns, told her mother she would never again play with matches.

These horror stories are a common occurrence. Parents who would never present a stranger with the keys to the family car often will put their trust in a relatively unknown or unqualified sitter to care for their children.

Sitters should be known to parents and, preferably, should be trained to handle emergencies and have some basic knowledge of first aid. A sitter who is new to the household should be introduced to the children while they are awake and, of course, to the family dog.

The sitter should be taken on a tour of the house and shown escape routes and problem doors or windows. She or he should be provided with a copy of the family's fire escape plan and a flashlight for use in an emergency.

Written instructions should be left with the sitter. These should include such details as where the parents can be reached in the event of any emergency, the name and phone number of the family doctor and the local hospital, and the name and phone number of a close friend or neighbor of the family who also might be notified in case of an emergency.

Local police and fire department phone numbers also should be left with the sitter. The sitter should be shown the location of the family first-aid kit and, instructed to use it for minor cuts or bruises suffered by the youngsters. Other instructions should require that the sitter walk through the house at least once every hour to make certain that all is well.

Other oral or written instructions should include:

- Accompanying children to the kitchen, stopping them from going to basement and utility areas, and keeping them away from heaters and matches.
- Keeping the radio, television, or stereo system turned low in order to hear any unusual sounds from the children.
- Permitting no social visits or any gathering of the sitter's friends in the home.
- Withholding permission to use heating or electrical equipment unless instruction has been provided.
- Limiting the number and length of the sitter's phone calls so that parents can call home to learn if all is well.

## The Elderly and Fires

Along with our concern for the safety of the young is our concern for the safety and well-being of the elderly. A recent study of accidents that happen to the elderly reported that persons sixty-five or older suffer 23 percent of accidental deaths. In recent years accidents involving the elderly averaged 25,000 annually. Nearly 2000 died in fires. The National Safety Council reported that people under age sixty with burns over 25 percent of the body have a 90 percent chance to recover. Those over age sixty have only a 60 percent chance to survive.

The elderly are more prone to accidents because the aging process often affects vision, hearing, mobility, and the ability to make quick judgments. Any of these weaknesses can lead to more injuries, especially those from fires. Older people who live alone often fall victim to kitchen fires while preparing their meals.

Smokers among the elderly are often fire victims, especially from burning matches that fall from unsteady hands. Smoking in bed is a dangerous habit for anyone, particularly the elderly, who often drowse or drop off to sleep without extinguishing the cigarette. Some older people require the occasional use of oxygen, and smokers among them have been known to light up in the vicinity of such equipment, contributing to an explosion or fire that burns faster and hotter when fueled by an oxygen environment.

Fire! Prevention, Protection, Escape

The most recent data about those sixty-five years of age or older indicate that this age group accounts for nearly 15 percent of victims of flammable fabric burns. Most fatalities among the elderly attributed to fabric burns have occurred in the kitchen. The bedroom is the second most frequent place where fatal fires have happened, followed by incidents in the living room and out in the yard.

The National Institute on Aging advises the elderly to prevent fires by adhering to these simple instructions:

- Never smoke in bed or when reclining in a chair or while tired and drowsy.
- Loose-fitting clothing with full sleeves, such as pajamas, bathrobes, or nightgowns, should never be worn while cooking.
- The hot-water heater and faucet thermostats should be set well below scalding temperatures.
- Dress in nonflammable clothing or wear clothing treated with a permanent flame-retardant finish.
- Always use several electrical outlets to avoid overloading circuits, which might cause a short circuit.

Finally, if an elderly person is to enter a boarding home or an apartment complex that serves the aged, carefully inspect the premises. Have smoke detectors been installed along with a sprinkler system, and are fire extinguishers handily available? Or, preferably, is there a fully automatic fire alarm system? Has a viable fire escape plan been developed at the residence, and is the escape plan regularly rehearsed by all occupants?

## ✔ Fire Escape Plan Checklist for the Young and Old

☐ Are all family members including young children checked out in the escape plan procedures?
☐ Has each babysitter been asked to study the family's fire escape plan?
☐ Have emergency phone numbers been left with the sitter? What about other safety measures?
☐ Are other lists of "can and cannot do" rules posted for all to read?

## Chapter 8

# Hotels, Motels, and High-Rise Buildings

America is a nation of travelers whether they cross the wide expanse of the United States or visit other countries. Many Americans travel regularly on business; others visit friends and relatives or just tour the country, often stopping for the night at a hotel or motel.

The threat of a fire follows us when we are on the move; therefore we have to be on our guard. Hotel and motel fires have made headlines for years. These fires have taken the lives of people who just didn't know what to do in a fire emergency. Many of them would be alive today if they had been prepared for the possibility of a fire.

Many hotel and motel fires go unreported. Most hotel managers are concerned about the negative publicity and loss of business that would follow the reporting of a fire.

Many hotel fires got in the headlines over the years because of the number of deaths involved. Or because prominent people were present. Most fatalities were caused by toxic gases carried in the thick smoke that filled several floors.

In a 20-story Las Vegas hotel, a fire broke out while entertainer Andy Williams was backstage preparing to go on with his act. The actress Juliet Prowse was performing at the time. Her act was interrupted by Dick Lane, another performer, who announced over the dining room's loudspeaker system that "there is an emergency. Please leave immediately and be calm." Among the injured who suffered from smoke inhalation was singer Natalie Cole.

In hotel or motel fires, guests often panic and do foolish things, such as stampeding down a smoke-filled stairway only to succumb to toxic gases created by the fire. Or they dash to an elevator in their flight to "safety," only to discover that the elevators aren't working. Others have tied up the hotel's telephone switchboard by trying to keep an open line. And there are those who have attempted to escape to safety by the most unsafe means: They have tried to lower themselves a perilous distance on bedsheets tied together or torn into strips knotted into a flimsy rope. Others have tried to jump or climb to safety. The will to survive is basic.

Emergencies bring out the best and worst in people. This chapter offers some suggestions that will bring out the best in those who heed its advice.

## Travelers' Checklist

If you are a regular business traveler—or, for that matter, an occasional traveler—it's wise to make up a standard checklist. The first part of your checklist is informational. It covers your trip: destination, method of travel, hotel accommodations, and length of stay.

Perhaps you are too busy to check out the hotel at which you will be staying during your business trip. If so, your secretary can obtain the answers to these questions:

- How many floors has the hotel?
- What kind of fire protection does the hotel provide?
- When did the hotel have its last fire department inspection?
- When was the last fire drill, and how many fire drills are held annually for the staff and guests?

These are optional questions; you don't have to be so specific. But well-managed hotels will provide answers to these questions. A broad general question you might ask is, What fire precaution and safety measures does the hotel provide?

You may be one of the many guests who prefer rooms below the ninth floor in a hotel. Another point of information you might seek is: Are the hotel's rooms and corridors fixed with smoke or fire detectors and sprinklers?

It's also important to know about fire drills. Hotels that hold fire drills during the course of a year usually have well-trained employees who know what to do in the event of fire, and such people are less likely to panic.

Pretravel requirements are important to your on-the-road fire safety planning. A few necessary items that take up little suitcase space could make all the difference in the event of a fire:

- A flashlight. And be sure to check the batteries every three months.
- A painter's mask. Hardware stores sell this inexpensive, lightweight item that will permit you to keep both hands free if you have to move about a smoke-filled area.
- A roll of carpet tape. Also a hardware store item, the tape can be used to seal a hotel room door against smoke.
- A small, portable, battery-operated smoke detector. This can be

### Fire! Prevention, Protection, Escape

mounted on top of the door, attached to the wall, or hung from a doorknob if your room has no smoke or fire detector.
- A small first-aid kit.

Don't be embarrassed by fire safety precautions. Many travelers are doing the same thing today. They are asking questions about the fire protection and safety features of hotels before they make their reservations.

There are two other questions that you can answer after you have gotten your room:

1. What is the location of your room in relation to fire exits?
2. Is there a floor plan with the fire exits properly marked?

Some well-run hotels instruct bellhops to mention fire prevention and safety features after they escort the guest to his or her room.

A few minutes of time spent in your hotel room taking the following precautions could save your life:

- *Check the windows.* Are they sealed closed, or do they open? Look out the windows and note if there is a ledge outside the window or a roof or parapet one, two, or three floors below.
- *Check the bathroom.* Note if there is an exhaust system typical of those used in a windowless room.
- *Check the door lock.* Note the kind of lock on the hotel room door and fit in the key one, two, or three times to become familiar with how it works and the direction in which the key turns.
- *Check the corridors.* After checking the room for a smoke detector and the location of a fire alarm, if one has been installed, leave the room for a few minutes and walk the length of the corridor in both directions. Note any obstructions between the room and fire exits, and the number of doors between the room and the nearest fire exit.
- *Check for fire exits.* Note the location of any fire exits in the hotel corridor and their location in reference to your room.
- *Check your safety equipment.* Always place your room key and flashlight on the bedside night table, and keep other safety equipment on a nearby dresser before retiring for the night. The majority of hotel fires usually occur at night or in the early hours of the morning.

The following rules apply in well-run hotels:

- If you smell smoke or the smoke alarm sounds, notify the hotel switchboard operator. The hotel staff will check it out. If there is a fire, the hotel staff will immediately call the fire department and at the same time see that guests are helped to safety.
- If awakened by a fire at night, grab your room key and flashlight and move quickly to the door. If there is any evidence of a

## Hotels, Motels, and High Rise Buildings

smoke-filled room, roll out of bed onto the floor. Keep low and crawl to the door with flashlight and room key in hand.

- Check the door for heat with the palm of your hand. Slide your hand from the top down, as you would at home if a fire was anticipated on the other side of a closed bedroom door (see Chapter 6). Note that some fire escape rules for the home can be used in a hotel or any high-rise building.
- If there is smoke or flame on the other side of the hotel room door, wet towels and/or that emergency carpet tape you brought along can seal cracks around the door.
- Don't forget that painter's mask for filtering smoke and breathing more comfortably. The mask permits the use of both hands to carry out safety and escape recommendations.
- Fill the bathtub with water to wet down the door, walls, or floor if they get too hot and the room cannot be evacuated. Water pressure often disappears when a hotel fire breaks out. Immediately filling the tub is a recommended precaution. And for those who don't carry a safety kit when they travel, a wet towel held over mouth and nose while crouched low will help filter out smoke irritants.
- If forced by circumstances to remain in the room, raise the bottom window, and lower the top window, approximately 4 inches if the hotel room has this kind of window. Air can come in from the lower window, and smoke will dissipate through the upper opening.

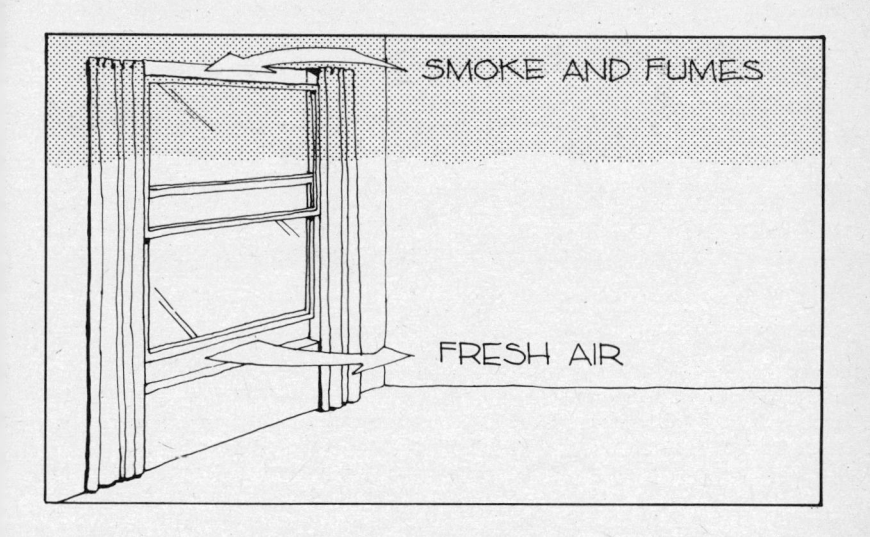

Fire! Prevention, Protection, Escape

- Hang a sheet or some brightly colored article of clothing out the window as a signal to firefighters that the room is occupied.
- If escaping from the room through a smoke-filled corridor is an option, don't forget to take the door key and flashlight along. Keep low and remain close to the corridor wall. Count the doors on the way to the stairs or to another fire exit so that you will know where to go if you have to retreat. Once at the staircase, grip the handrail for guidance and protection against being knocked down by panicky guests.
- If the smoke becomes too dense on lower floors, retreat up the stairs to the roof if this is practical. Prop open the roof door to vent the stairwell and prevent the door from locking. Also, move to windward side; that's the side away from the direction of the smoke.
- *A final warning:* Never use the elevator to escape from a hotel or high-rise building. During a fire, elevators rarely go to the floor to which they are summoned. They often remain at one floor with doors jammed open. Smoke can "break" the electric eye beam that controls elevator doors and stop them from ascending or descending. The bodies of many victims of hotel or high-rise fires are found in elevators stalled on a smoke-filled level. Most people who have died in such fires succumbed to smoke and toxic poisons, not to flames.

### ✓ Fire Safety Checklist for Hotels, Motels, and High-Rise Buildings

☐ Have you prepared a travel checklist?

☐ Do you have an emergency kit with flashlight, carpet tape, painter's mask, and a small, portable smoke detector?

☐ Did you take those extra minutes after registering to look for the hotel room's emergency fire instructions and walk the corridor to check out the fire alarm, stairwell, or other fire exit?

Chapter 9

# Seasonal Fires

There is a season to live—and a season to die. The incidence of fires in the United States also follows the seasons. These newspaper headlines say it all:

> Kerosene Burner Explosion Kills Two Children
> Electric Heater Shorts, Four Succumb
> Clogged Fireplace Flue Destroys Home
> Kitchen Blast Takes Five Lives
> Three Die in Christmas Tree Fire

These headlines all represent cold-weather fires. They were taken from newspapers in New York City; Forest Heights, Maryland; Vinita, Oklahoma; and York, Pennsylvania. Such tragedies are often referred to as "seasonal fires" by fire safety experts, for they occur with the regularity of the seasons year after year.

In the Pennsylvania fire, two youngsters who were left unsupervised perished when one of them tried to refill a kerosene heater in his bedroom before going to sleep. The child tried to lift a five-gallon can of kerosene, as he had seen his parents do, but the can was too heavy, and he spilled some of the fuel on the hot heater. He had inadvertently created a vapor fuse. Flames flared up from the heater and reached into the kerosene can, which exploded in his hands. The blast killed him and his younger sister.

An improperly installed wood-burning stove also took its toll. A buildup of creosote in a chimney that had not been cleaned for three years caused a fire that destroyed a home when the fireplace was used. Creosote is caused by the combustion process of burning wood, which never burns completely. Unburned gases in the flue build up creosote on the cooler inner surfaces of the chimney and flues. As these deposits are warmed, they form a sticky, tarlike substance that will ignite when very hot and will cause a chimney fire that is similar to a grease fire in a kitchen stove ventilation system.

### Fire! Prevention, Protection, Escape

A short circuit in a portable electric space heater started a fire when a very dry and cracked electric cord caused sparks that landed on drapes and set them afire. In another incident, Christmas tree lights were stored in an attic for more than six years. The wire insulation dried up and cracked during the hot summer months. When the lights were used, sparks from the ruptured insulation set the tree on fire.

Each of these fires could have been prevented if care had been taken.

## Wood-burning Stoves

Wood-burning stoves can be useful, cozy, and even picturesque. But they are very dangerous if not properly installed, maintained, and operated.

Wood-burning stoves come in different sizes. A small stove is most often designed to warm a smaller room; a larger stove obviously gives off more heat and can warm a larger room. As a rule, the stoves sold today are for heating individual rooms and not entire homes. It's important that any wood-burning stove purchased, or any heating system for that matter, be accompanied by installation and operating instructions.

Some people purchase used wood-burning stoves. A word of advice: carefully inspect any used stove for broken parts or cracks. These damaged parts of the stove can make the stove unsafe to use.

The Federal Trade Commission (FTC) recommends that safety be the main concern of purchasers of wood-burning stoves. The National Bureau of Standards warns that unless wood-burning stoves are installed properly and used carefully, "they can cause disastrous fires." A Cornell University study by environmental scientists warns that wood fires can give off pollutants that can be dangerous in homes without adequate ventilation.

Fire prevention for wood-burning stoves begins by taking the time to do some homework. First, consult with your local fire department—it's free of charge. Then:

- Check your municipality's local building code for the wood-burning stove models that are acceptable.
- Evaluate carefully all advertisements about wood-burning stoves (or any heating device) and how well they perform. Remember, advertising is meant to generate sales.
- Check the "net efficiency" rating. The FTC says this rating is the indicator of how much heat actually is expelled into the room by a particular model wood-burning stove. The "combustion efficiency" indicates how well the particular stove converts wood to heat.
- Purchase only a cast-iron or steel stove that bears the approval seal of a recognized testing laboratory.

## Seasonal Fires

Although local regulations vary, wood-burning stoves should be installed at least 36 inches from combustible areas such as ceilings and walls, and from furniture, carpeting, and drapes. The stove should always be mounted on a floor protected by a stove mat or other certified noncombustible floor covering that extends at least 18 inches beyond the stove's outer edges.

Wood-burning stoves require a simple exhaust—a stovepipe to carry the fumes out of the house. The stovepipe can be connected directly to the chimney or, if there is no alternative, through a wall to the outside of the building.

The stovepipe attached to a wood-burning stove also becomes a conveyer of heat. Hot exhaust fumes travel from the stove to the chimney or directly from the room in which the stove is used to the outdoors. Protective materials must be used to insulate that part of the stovepipe that passes through a wall to outdoors or to a chimney.

Improperly vented wood-burning stove chimneys can produce deadly amounts of carbon monoxide as well as other deadly gases such as carbon dioxide and oxides of nitrogen, sulfur, and other toxic pollutants.

Always use wood as the basic fuel for a wood-burning stove. Never attempt to start a fire in the stove by using gasoline or a charcoal fuel. Also guard against overheating; this is the most common problem with this kind of stove. Use seasoned or dry wood; it burns better and provides more heat value.

Heavier hardwoods such as ash, beech, maple, and oak provide longer-lasting fires and may be mixed with softwoods, which start a fire more easily. Store wood in a sheltered place to keep it dry and prevent or minimize rot. Wood should be stored for six months before use.

One caution when using a wood-burning stove: Never burn wood that has been treated with paint or another chemical base. Wood discarded from construction, and therefore treated, burns unevenly and produces dangerous toxic gases. Paper "logs" should not be used in a wood-burning stove.

Never attempt to burn vacuum cleaner bags filled with dust. The confined dust in the bag tends to explode when burned.

Wood-burning stoves are meant to be fueled with seasoned, dry wood. Anything else invites a problem or danger.

Here are some further tips on using a wood-burning stove:

- Make sure that all stove ashes are thoroughly cooled before disposal. Use a metal container with a tight-fitting lid for disposal of ashes.
- Always keep children away from wood-burning stoves or any other supplementary heating system to prevent burns caused by touching a hot surface.
- Do not dry cloths or clothing by hanging too close to the hot wood-burning stove.

Fire! Prevention, Protection, Escape

## Kerosene Heaters

Another kind of room heater is fueled by kerosene and has become increasingly popular because of the high cost of home heating oil and electricity. Unfortunately, kerosene heaters are fire hazards. Ignorance and carelessness in using kerosene heaters have caused many tragic accidents, especially in inner-city urban areas.

A number of states and many cities have declared the use of kerosene heaters illegal because of the number of deaths attributed to their misuse. Entire families have died of asphyxiation in urban tenement buildings where residents use the heaters to heat rooms without windows or with windows shut tight against the cold.

Any attempt to heat rooms this way can use up the available oxygen and create deadly, odorless, and invisible carbon monoxide fumes. The same danger holds true for using gas stoves to heat the kitchen or other rooms.

Newer models of kerosene heaters have cutoff switches that automatically turn off the appliance if it's knocked over. But this is no protection against an accidental fire caused by spilled kerosene that comes in contact with a hot surface.

## Electric Heaters

In recent years efficient portable electric heaters have become popular. They come in many sizes, can easily be carried from one room to another, and plug into any electric outlet. A feature of the newer models is an automatic cutoff switch that shuts off the heater if it's accidentally knocked over.

Here are some tips on using electric heaters:

- Never use an electric heater in a bathroom. Water or dampness can provide an electric shock.
- Check wiring in the home to make sure that it will take the extra power needed to operate an electric heater.
- Portable heaters should be kept at least 3 feet away from draperies and other materials that might burn.

## Fireplaces

Another winter fire hazard in the home, and one commonly associated with rural and suburban areas, is the brick or stone fireplace. A fireplace brings to mind a warm feeling, romance, and a cozy glow from flickering flames. Fireplaces are enjoyable to use, but they are not toys.

Fireplaces can be dangerous, as fire department reports testify. First-time users should make certain that the fireplace was built to be used and is not just a living room or den decoration. A usable fireplace is properly constructed with the necessary interior lining and clearances that guard against an accidental fire.

The home fireplace should be checked annually to make sure the vent and chimney operate properly. One danger is the buildup of creosote inside the chimney, which can cause a chimney fire. The services of a professional chimney sweep, or cleaner, help prevent this kind of fire. Also, during the cold season, chimneys should be inspected regularly.

Here are some things to watch for when using fireplaces:

- A fireplace requires a good draft to prevent the accumulation of smoke and toxic gases.
- A protective screen should be placed across the fireplace to prevent sparks from popping out and setting fire to carpets or pillows or kindling. Carpets and pillows should be at least five feet away from the front of the fireplace. Fireplace sparks from burning wood have been known to land more than five feet from the fireplace; thus the need for a screen.
- Chimneys should have spark arresters installed to prevent sparks from touching off fires on wood shingles or in leaves that accumulate in rainwater runoff gutters.
- The fireplace damper should never be closed while hot ashes remain in the fireplace. A closed damper can cause heat to build up and create a sudden flare-up of hot ashes in a very low-level "explosion" that can scatter sparks and burning embers.
- A fireplace damper closed while ashes are hot prevents deadly carbon monoxide fumes from escaping.

An added safety precaution during the winter months is to remove from the fireplace mantel any materials that can burn. Also, do not use flammable liquids to start a fire or restart a cold fire. Coal, plastic foam packaging, or charcoal should never be used in the fireplace; they produce highly poisonous gases. Children should be taught to keep their distance from the fireplace. Clothing can ignite if youngsters get too close to a roaring fire.

Taking these simple fireplace precautions can provide comfort, tranquillity, and safety.

## Christmas Hazards

Many Christmas fires can be prevented by observing a few safety rules. For example, if you feel you must have a natural, or "real," tree, the safest kind is one that has been dug up with its roots in soil and then bagged. A daily light watering of the bagged roots and soil will keep the tree fresh and alive. After Christmas is over, it can be planted outside to enhance the home's landscaping.

If you choose a cut tree, which is more of a fire hazard, pick a tree that has been freshly cut check by shaking the tree. Once installed, the base of the tree should be kept in water, and the water level should be checked daily and replenished when necessary. Water will ensure a greener tree and one with enough moisture in its branches to reduce its flammability. I most highly recommend an artificial Christmas tree that looks like the real thing and is many times more safe to use. It also can be used for many years. Nevertheless, artificial trees are of many kinds, and some are made of plastic, wood, metal, and other materials. So purchase an artificial tree that bears the seal of a known testing laboratory.

Christmas trees should be kept no longer than 15 days. Immediately remove the tree from the home once it has been taken down. Do not erect Christmas trees near doors, stairways, or close by any heat source. Setting a tree close to a radiator or fireplace can cause rapid drying and consequently make the tree more flammable.

Lighted candles should never be used to decorate a tree nor be permitted nearby. All electric decorations should be carefully checked each year, and new ones must bear a testing laboratory's warranty. Such decorations should always be turned off before going to bed or when leaving the home unattended. It's much safer that way. Only noncombustible decorations should be used. Care should be exercised with sprays or with the tinsel and "angel hair" decorations that are in vogue.

Tests have shown that the application of fire-retardant sprays is not always reliable. Use fire-retardant sprays that bear a testing laboratory's seal. Other electrical devices, such as toys, trains, or spark-generating equipment, should be kept away from holiday trees.

House pets also should be kept away from holiday trees. They have been known to topple a Christmas tree. The tree can then catch fire when it comes in contact with "hot" electrical elements.

Christmas trees come in different materials and require special care. For example, the all-metal trees should never be decorated with electric ornaments. They can cause a short circuit if a defective wire comes in contact with the tree's metal construction. Another tree that must be inspected annually is the artificial model decorated with built-in miniature lights. Any built-in features require more care.

A safe rule for the Christmas season is to purchase a basic tree that is artificial and bears the seal of reputable testing laboratory. The payoff is peace of mind

## Home Workshops

Fire prevention, protection, and safety should be year-around watchwords. During the winter months, however, more time is spent indoors. And free time is often spent in the "home workshop," which may be a corner of the cellar or the garage. Although other hazards in the home workshop come from the misuse of tools or machinery, the most life-threatening danger is fire.

Every home workshop holds a fire hazard of one kind or another, but the possibility of a fire can be reduced by remembering these basic facts:

- Deposits of wood shavings, sawdust, and ground metal are highly flammable when they come in contact with fire. A spark from an electric switch has been known to set off a flash fire when finely ground residue of workshop "dust" is in the air. For example, a flame or an electric spark can cause workshop sawdust to explode or burn rapidly. Metal dust holds these same dangerous characteristics.
- Good home workshop housecleaning habits can help prevent fires. Clean up the residue from grinding, sawing, sanding, and planing immediately after work is completed.
- Fire prevention storage rules in home workshops apply in the same way as in the residential part of the home. Flammable liquids should be kept in tightly closed containers and stored in well-ventilated areas away from matches, pilot lights, sparks, or hot surfaces.
- Vapors from different liquids used in the workshop are usually heavier than air and travel close to the floor. A spark from a cigarette ash has been known to ignite a vapor trail and set off a flash fire.
- Ignition of sawdust and vapors is an ever-present danger that *Workbench* magazine contends has been the cause of many home workshop fires started by electric space heaters. The magazine also recommends that all electrical outlets in home workshops be well grounded and accept three-pin male plugs attached to heavy-duty extension cords. "Improperly maintained hot water heaters, furnaces, motors or poorly functioning electric tools are all potential sources for stray sparks," warned *Workbench* in 1971. "You should carefully monitor their operation and maintenance to prevent a fire."

## Tools and Equipment

All states are faced with fire prevention problems, many of which are seasonal. When April rolls around, for example, residents in the northern half of the nation begin to move their activities out-of-doors. Lawn mowers are cleaned and adjusted along with other outdoor power-driven lawn tools. Campers and recreation vehicles are cleaned, and camping equipment is hung out to air. Boats are sanded and painted and otherwise readied for the water.

All these activities mean added fire possibilities, and so they require attention to detail and safety. In the fall, rake leaves and pack them into large plastic bags. Many suburban municipalities ban the burning of leaves. Some rural areas have few regulations against burning leaves, but watch for the posting of the fire warden's notice that it's either too dry for burning or that there is enough moisture on the ground and in the trees to proceed with burning leaves. Farmers who wish to burn off dried-out crop debris should do so under the supervision of the local fire department.

In suburban locales north of the Sunbelt, spring begins by raking leaves and cleaning lawn tools, many of which are gasoline powered. The threat of accidents and fires can be lessened considerably by not refueling the lawnmower when the motor is running or when the mower is inside a tool shed or garage with limited ventilation. The same rule holds true for other gasoline-powered outdoor home maintenance tools or equipment, such as chain saws, snow blowers, small tractors, and yard vehicles.

Here are some other safety tips on tools and equipment:

- Check the condition of lawn mower mufflers. Hot gases from defective mufflers can ignite dry grass. Spark arresters should be used in areas where dry grass is common.
- Electric-powered lawn mowers, saws, and other yard tools must be carefully checked before use. Cut or frayed extension power cords should be repaired immediately.
- Use only extension cords and plugs—and equipment—that carry a testing laboratory's certification.

## Home Barbecues

The home barbecue is an American institution that requires fire safety precautions. Around the home, fire prevention extends to charcoal grills as well as to lawn and yard equipment. Summer can be a fun season, but only if it remains accident-free and fire-safe. Some of the precautions to be taken at the charcoal grill are the same as those to be observed in the kitchen.

- Loose clothing should never be worn near the barbecue grill. Flaming grease can ignite clothing.
- Use only testing laboratory-recommended fire-starter fluids. After use, return the container to its proper storage place away from the grill. Do not add charcoal starter fluid to the fire after it has started. A vapor trail leading to the container can cause an explosion and injury. Starter-fluid containers should be purchased with childproof caps and kept away from children.
- Keep the area around the grill free of twigs, dried leaves, paper cups, plates, and napkins.
- Trim the limbs of trees near the barbecue area to prevent setting fire from a flare-up of fuel or grease.
- Keep a box of baking soda at hand to smother an out-of-control fire.

# Seasonal Fires

- Never use charcoal barbecues in an enclosed space. Charcoal briquettes give off quantities of carbon monoxide, which can be dangerous.
- Use caution when disposing of charcoal grill ashes. If not thoroughly extinguished, they may contain live coals or hot embers that can start a fire.

In recent years many charcoal grills have been replaced by more sophisticated home barbecues fueled by liquefied petroleum gas (LPG) contained under pressure in a steel cylinder. The contents of an LPG cylinder, vaporized in a confined area, have the explosive force of several sticks of dynamite. The wise user of LPG will be aware of the danger and the precautions that must be taken to avoid accidents.

Never ignite a cooking or heating unit that uses LPG until the following precautions have been taken:

- Read the instructions carefully before connecting the LPG cylinder to the grill; use the proper-size wrench; and make certain that all connections are tight. *Remember that fittings on flammable gas cylinders have left-hand threads that are tightened by turning in a counterclockwise direction.*
- Make certain all connections are tight by applying a soapy solution to detect leaks. If bubbles appear, tighten further until they stop.
- Prevent grease from falling on the hose or cylinder.
- Do not permit children to use the LPG barbecue.
- Never store the portable LPG grill inside any structure or under a staircase, porch, or balcony. Cylinders should be stored outdoors in a shaded, cool area.
- Do not transport LPG cylinders in an automobile trunk. Full cylinders *always* must be transported in an upright position on the floor of the vehicle and with windows open if it's an automobile. Remove the cylinder as soon as possible after arrival at your destination; *never* leave a full cylinder in a parked vehicle.

## Fireworks

The seasons offer a variety of activities. But an awareness of seasonal fire dangers can prevent accidents, especially around the Fourth of July. In most states and municipalities, fireworks are outlawed. Yet the bootleg sale of fireworks continues, and young and old suffer burns and other injuries.

Illegal fireworks, whether made in America or imported from other parts of the world, are manufactured *without* safety standards. Exploding firecrackers of varying sizes pose the greatest danger. Some noisemakers explode prematurely in the hands of a youngster or adult,

### Fire! Prevention, Protection, Escape

and fireworks have been known to blow off fingers, blind, and otherwise mutilate victims. Premature explosions of fireworks are caused by rapidly burning fuses that give little time to move away from a disabling blast or throw the firecracker to a safe distance.

Skyrockets also have been known to cause fires. They can fall on rooftops or in wooded areas, which is especially dangerous during the dry season.

Other dangerous fireworks are pinwheels, fountains, snakes, and sparklers—all of which provide color along with danger. People who play with fireworks are rarely safe from fire injuries. They set off fireworks without a thought about the danger to themselves or others.

One of the best ways to usher in the Fourth of July is to watch a fireworks display assembled by experts for the enjoyment of all. And even experts have been maimed or killed by firecrackers.

## Halloween

Another seasonal event that's hazardous to children is Halloween, the final night in October, when youngsters of all ages take to their local neighborhoods to "trick or treat." In recent years, parents have become aware of the need to warn their children about not eating any treats until Mom or Dad has had an opportunity to carefully examine the evening's collection of candy.

Mothers also are very careful about providing costumes that reflect headlights. In fact, many parents are so careful that often one of them escorts the youngsters from house to house in the neighborhood. You might think that parents have covered every safety eventuality. Unfortunately, this isn't the case.

Pumpkins—real or artifical—can cause pain and injury. What happens is that all too often, flimsy paper and flammable, lightweight cloth costumes are purchased for youngsters. The burning candle in the pumpkin carried by a child can throw off sparks on a chilly and windy night. Year after year, the sparks from Halloween pumpkins transform a night of fun into a night of horror when a child's clothing catches fire.

## ✔ Fire Safety Checklist for Seasonal Fires

☐ Is your wood-burning stove installed properly?
☐ Do you thoroughly cool stove ashes before you dispose of them?
☐ Has your fireplace chimney been cleaned recently?
☐ Are you particularly fire-prevention-minded at Christmas time?
☐ Do you keep a clean home workshop?
☐ Are you careful with your gasoline or electrically powered outdoor tools?
☐ Do you keep a fire extinguisher or a box of baking soda nearby when you barbecue outdoors?
☐ Is your LPG cylinder kept in a cool, shaded place?
☐ Are you a fireworks watcher instead of a user?
☐ Are your children kept away from Halloween pumpkins illuminated by candles?

## Chapter 10

# Fire Safety and Recreation

## Boating

Boating is a growing American pastime that requires careful attention to fire safety. In fact, there are procedures and rules that must be followed carefully.

Rules must be followed when the boat is afloat and when it is out of the water. There is no special time for boating enthusiasts to observe fire safety practices, advises the National Fire Safety Protection Association. Marine fire safety and prevention should apply to all boat maintenance activities, whether carried out in a backyard, a boatyard, or on the water.

Special attention should always be paid to such chores as paint removal, scraping, sanding, varnishing, cutting, and welding. Some helpful fire prevention hints follow:

- Never accumulate oily rags. Promptly remove sawdust and wood shavings. Properly store leftover paint and painting supplies before departing the work area.
- Remove paint by scraping, hand and machine sanding, or wire brushing, using only nonflammable paint remover. In the event that a flammable paint remover is used, the work must be done outdoors.
- Strictly observe the No Smoking signs while painting boat interiors or working in enclosed areas. Always maintain adequate ventilation, and never operate equipment that might produce sparks. Do not allow any kind of open flame near a boat.
- Get professional help for welding. If brazing, soldering, or metal cutting, insist on the highest professional safety standards when doing such work on a boat.
- Always keep a fire extinguisher or garden hose at hand, set up for immediate use.

## Fire Safety and Recreation

Fire safety habits on a boat can be lifesavers once the craft is in the water. Marine safety regulations established by the U.S. Coast Guard are practiced by boaters who have taken this government agency's sailing and safety courses. Unfortunately, too many amateur sailors endanger others by their ignorance of fire prevention and safety rules that apply to boats.

Carelessness during fueling and on-board boating operations can turn pleasure to tragedy. For example, when refueling their boats at night, experienced sailors do so only under well-lighted conditions. There is no smoking aboard or on the dock during the operation. Prior to refueling, all engines, motors, fans, and heating devices are shut down; all flames are extinguished; and all ports, windows, doors, and hatches are "secured" so that the boat is completely "dead."

Fuel tanks are *never* filled to capacity, to permit space in the tank for fuel expansion. During refueling, the delivery nozzle is kept in contact at all times with the fill pipe, to prevent a spark. The nozzle is firmly held in place before fuel delivery begins and is removed only when such delivery has ended. Spills during refueling are wiped dry as soon as the procedure has ended.

When the portable fuel tanks of outboard motors are filled on board, only approved metal containers with proper capping devices should be used. Gasoline should never be carried on board in glass jars, plastic containers, or open buckets and cans.

The experienced boater sniffs the air before starting an inboard engine. He inpects bilges for fuel leakage or odors, and allows enough time to ventilate areas below deck until the odors disappear. Bilges are always kept free of waste, rags, paper, and other combustibles, including lubricant spills and fuel vapors.

Many boats have small galleys. These galleys vary with the size of the boat and run from simple to elaborate. Galley stoves should be designed for use aboard a boat and should carry the approval of a certified testing laboratory.

Some galley stoves are powered by electricity. Others use alcohol or liquefied petroleum gas (LPG). Both propane and butane are highly flammable, and are dangerous if misused. Gasoline is not a safe fuel for use in a marine stove; the U.S. Coast Guard warns that it should *never* be used.

The Coast Guard recommends that galley stoves be fastened securely. Adequate ventilation must be provided to prevent excess temperatures from building up in the galley area. All woodwork immediately surrounding a stove must be sheathed with a fireproof material such as asbestos board, covered with sheet metal.

Fuel should be supplied to the burners by either a gravity or pressure-flow system. Refueling should be permitted while the stove is in use *only* if the supply tank is remote from the galley and the burners. Pressure tanks should have suitable gauges and relief valves.

Fire! Prevention, Protection, Escape

Water can be used to extinguish burning alcohol. An ABC extinguisher can stop a small LPG fire.

For fire protection, boats should carry the approved fire extinguishers recommended for the class of the boat. Fire extinguishers should be kept close at hand and located just outside the compartments to be protected. On-board fire extinguishers should be fully charged and in first-rate working condition at all times.

Finally, never leave fire safety precautions behind when you shove off in your boat for fun on the water. After all, it's your life and the lives of your family and friends that are at stake.

## Before Your Vacation

If you're about to leave on vacation, home fire safety precautions should be taken before your departure, along with the normal procedures of canceling deliveries of newspapers, milk, mail, and other items. A neighbor should be available to keep an eye on your home and should perhaps be provided with a key so that the inside of the house can be inspected regularly.

There are other important precautions you should take. They relate to fire prevention.

- Disconnect or turn off the stove and all electric appliances.
- Unplug TV sets, radios, and stereo players.
- Empty all trash from the house and outside trash cans, to lessen the chance of a fire occuring.

If you are traveling by auto, you have spent time tuning up your vehicle. The luggage is loaded, but if one of the family asks about carrying a can of gasoline in the trunk so that you won't run out of gas, carefully explain that this is a lethal practice. Gasoline should *never* be stored or carried in the trunk. A rear-end collision, which could rupture a gasoline can, could create a spark and an explosion. Cars have been known to explode into fiery infernos in such a short time that no one in the automobile was able to escape.

Once on the way, be careful. Don't discard cigarette or cigar butts on the side of the road from your car. Discarded materials have started roadside fires during dry spells. It doesn't take that much effort to use your automobile ash tray. And carry an ABC or BC fire extinguisher in your car; have it close at hand for use in an emergency.

## Camping

If you plan on camping out, keep in mind that automobile exhaust pipes get very hot when the vehicle is in motion or idling. Parking on dry grass or in brush can start a fire beneath the auto. Another fire hazard can occur while camping: A burning tent can be just as dangerous as a house on fire.

## Fire Safety and Recreation

Tent fires are just one of the fire hazards at a campsite. In one camping fire, three adults and four children bedded down in a tent at a Connecticut camping ground. The tent flaps were zipped tight against a steady rain that had been falling. The heat from a gasoline pressure lantern soon became unbearable in the closed tent.

While the children slept, the adults conversed quietly. One of them struck a match to light a cigarette. There was a *whoosh* and an explosion inside the tent. Burning canvas fell on the occupants. The adults scrambled to safety, pulling one of the children with them. The other three children died when the tent was consumed by flames.

How could a camping tent burn so quickly? That was the question that had to be answered. Fire investigators learned that on the afternoon prior to the tragedy, the tent's owner had applied a water-repellent chemical to the canvas. The water-repellent container was labeled "flammable." The water repellent apparently had not dried sufficiently before the tent was used. Fire investigators theorized that a buildup of volatile vapors in the enclosed space was susceptible to any flame or spark, which would set off the fumes in a flash fire.

Some tents are flammable; they can ignite if erected near an open flame and can burn completely within minutes. About 80 percent of all camping tents are made of cotton, and the U.S. Consumer Product Safety Commission warns that waterproofing a tent sometimes increases flammability.

Here are some fire safety camping rules:

- Never use candles or gasoline pressure lanterns in a tent, and never set up a tent too close to a campfire.
- Always refuel any heat-producing equipment, such as a lantern or stove, at least 15 feet away from a tent. Store flammable liquids at least 30 feet away.
- Never cook inside a tent.
- Always pitch a tent away from a campfire and upwind so that sparks will not fall on the tent.
- Don't hang a bare electric light bulb where it may touch the tent fabric and burst into flame.
- Use tents that are manufactured of flame-resistant material.

Part of the fun of camping out is sitting around a campfire after a hearty meal, without any thought given to fire, fire prevention, and fire safety. But even the campfire can be dangerous unless some thought is given to preventing the fire from spreading.

Here are some campfire tips:

- A campfire should always be made in a designated area that is first cleaned of leaves and other debris to prevent the spread of flames.
- Don't build a bigger fire than you need, and never try to start a campfire with gasoline or a lighter fluid.

Fire! Prevention, Protection, Escape

- When finished with the campfire, drown the flames with water and stir the ashes.

A summer excursion like camping out can be ruined by sudden thunderstorms or, as they are sometimes called, "electrical storms." These noisy downpours, preceded by flashes of vivid lightning, occasionally cause fires and play havoc with home electrical systems. A direct hit by a bolt of lighting can fell a tree or kill a person; and those nearby can receive electric burns.

If a thunderstorm occurs while you are camping, get out of the water, stay away from wooded areas with high trees, and seek out a low spot such as a gully or depression in the landscape. Avoid metal pipes, umbrellas, telephone wires, and metal camping equipment. If you're at home when such a storm breaks, unplug all major appliances and avoid standing near metal heating ducts, radiators, and water pipes. Outside antennas have been known to draw lightning to the television set.

## Recreational Vehicles

As the cost of air and surface transportation increases, more Americans are purchasing recreational vehicles for vacation and leisure-time use. Sales of campers, motor homes, trailers, and vans increases each year. "People are determined to find a less expensive way to travel," notes the Recreational Motor Vehicles Association.

Some of these vehicles have been transformed into mobile homes on wheels, and the term "mobile home" has acquired another definition: a home set on its own wheels that can be transported to a site by a tractor-trailer truck.

The increase in the number of recreational vehicles (RVs) and leisure-time activities enjoyed by Americans permits them to "get away from it all." Unfortunately, they can't get away from the ever-present threat of fire. Some precautions that should be taken while on the road in an RV require a periodic check of all connections in gas-fired stoves and LPG cylinders.

- The fuel release on LPG cylinders should be shut tight while the RV is moving on the highway.
- Cooking in an RV while it's moving is dangerous. A sudden lurch or braking could cause a grease spill or cooking fire.
- Always fuel stoves or lanterns outside recreational vehicles to prevent an explosion caused by vapor accumulation from volatile fuels.
- Avoid accumulating and storing combustibles, such as empty grocery bags and newspapers or trash for disposal at "the next stop."
- Have an approved fire extinguisher on hand.

## Fire Safety and Recreation

The same rules that apply to careful automobile driving apply to drivers of RVs: never drive while tired or sleepy; and always pull over to the roadside to rest.

Mobile homes or, as they are often called by sellers, "manufactured homes," are usually primary homes in rural areas or second homes. In the 1940s and 1950s the typical mobile home was a small trailer hauled behind an auto. Later on, these trailers became longer and larger and were parked at trailer camps. In the 1960s mobile homes often were built to the same dimensions as large railroad freight cars or truck trailers, and the only "mobile" thing about them was moving them to a permanent homesite.

Upon delivery they were set on a cinderblock base and for all practical purposes were regarded as residential structures.

As the cost of conventional homes soared along with mortgage rates, less costly mobile homes became popular, especially for young couples who could not afford to purchase the three-bedroom-two-car-garage American dream.

Prior to 1974, many so-called mobile homes had little fire-resistant construction. Many of these homes are still in use. There are an estimated 25,000 fires in mobile homes each year, according to the U.S. Fire Administration (USFA). On the other hand, the U.S. Bureau of the Census notes that an estimated 11 million Americans live in mobile homes, and the number of such residents grows each year.

Residents of mobile homes should follow the fire prevention and safety measures recommended by the USFA National Fire Data Center.

- Install at least two smoke detectors in a mobile home. Locate them well away from the kitchen area to avoid false alarms triggered by cooking fumes. It's also suggested that the smoke detector be attached to an outside attention-getting device that sounds an alarm.
- As in a standard home, mobile home residents should prepare an appropriate fire escape plan and carry out periodic fire drills.
- Escape windows are critical in mobile homes; jalousie and awning-type windows should swing outward. Bedroom windows more than 36 inches above the floor are not good escape windows.
- Portable space heaters should not be refueled inside a mobile home; should not be used without proper venting; and should not be set up near window drapes.
- Heaters and stoves should be cleaned at least once a year by competent specialists.
- Fires in mobile homes can start in the fuse or circuit box when too many appliances are plugged into one electrical circuit. If lights blink or dim, motors slow down, or television and radio

volume change, turn off some of the appliances and check the fuse or circuit breakers usually located behind the master bedroom door or in a closet.
- Immediately call a licensed electrician if circuit breakers continuously trip, fuses blow, or acrid odors form.
- Install a "skirt" around the base of a mobile home to prevent leaves and other flammables from gathering under the structure and becoming a source of combustion during extremely hot or dry periods.

An additional fire protection precaution for a mobile home calls for at least three small ABC fire extinguishers. These should be placed in several strategic locations.

## Fire Safety Checklist for Boating, Camping, and Recreational Vehicles

☐ Do you follow strict fire safety rules for boats?

☐ Do you realize the danger of storing a gasoline can in your auto trunk?

☐ Is your camping tent fire-resistant?

☐ Are you careful with campfires? Do you pour water on the fire before leaving the campsite?

☐ Do you follow strict safety rules when using an RV?

☐ Is your mobile home protected by smoke detectors?

☐ Have you developed a fire escape plan for your mobile home?

☐ Is the area under your mobile home free of leaves and other combustible matter?

## Chapter 11

# Fighting a Home Fire

Fire!

Shouted or signaled, this warning can induce great fear. A fire can make us act irrationally. And irrational actions often cause the tragedies that make the headlines.

No two fire situations are alike. There is no pat answer to the question whether a person should try to fight a home fire or should evacuate the premises. Any response to an alarm that a fire has broken out depends on circumstances. But it's a critical decision. And if you make the wrong decision, it can cost you your life!

Always keep this warning in mind: Fire extinguishment is a dangerous business! It's a job for professionals.

There are no set rules for fighting a fire confined to its point of origin. Fire by its very nature is unpredictable. If you are present when a fire breaks out, any decision to take action requires split-second judgment. If you respond quickly and correctly, the odds may be in your favor that you will extinguish the fire successfully. If you choose to leave the building, especially if it's your home, there are other considerations—especially the safety of those you love.

A rule of thumb when a fire breaks out is that if you have the slightest doubt about your ability to extinguish the fire, evacuate the premises immediately. Let the professionals handle the situation. Simply put: If in doubt, get out!

With this statement in mind, here are some helpful hints to assist you in making a value judgment about whether or not to fight a fire:

- Did the fire start in your presence? Is it confined to its point of origin? If so, has it spread to other combustibles in the room of origin? You may find it possible to extinguish a confined blaze. If you make that choice, always keep your back to an exit so that you don't get trapped. Make sure your escape route is open.

### Fire! Prevention, Protection, Escape

- If the fire started in another location in the apartment, home, or workplace, do not try to find the fire and put it out. There are too many dangerous, unknown, and unpredictable factors. Evacuate the premises immediately and call the fire department.

A fire confined to its point of origin usually can be tracked. For example, if a fire started in a trash basket into which a burning cigarette from an ash tray was accidentally emptied, the chances are that the fire can be confined and extinguished. (Remember, always keep the door to your back in case you have to leave the premises quickly.)

On the other hand, if the wastebasket stands next to combustibles, such as drapes or a desk strewn with papers and other documents, there is every likelihood that the fire will touch off these items quickly and will spread. The spread of a fire beyond its point of origin is a warning that it usually cannot be extinguished by one person. In this case, get out of the building immediately and notify the fire department.

There are numerous incidents of people who have evacuated a building that's on fire only to rush back into the structure to save a pet or a valuable item. This is foolish—and dangerous. Many times, those who have rushed back into a burning building have been quickly overcome by toxic fumes and have died.

## Smoke Detectors

Different kinds of personal firefighting or fire safety equipment are available to the public. To begin with, one of the best lifesavers is the home smoke detector. It's credited by the National Fire Protection Association and the U.S. Fire Administration with saving many lives. Of course there are other, more sophisticated fire warning systems, but the simple, inexpensive smoke detector that gained acceptance in the early 1970s is widely used today.

National organizations like the NFPA and USFA highly recommend that every home have at least one smoke detector. Local fire departments also recommend their use, and in many cities the law requires that the landlord of an apartment building install a smoke detector in every apartment. Some insurance companies offer discounts to fire insurance clients who install smoke detectors in their homes or apartments.

The nation's home fire fatality rate has dropped significantly since smoke detectors appeared on the market. The findings of a study of home fires by the Federal Emergency Management Agency indicate the following:

- The chances of dying in a fire are reduced by up to 50 percent if smoke detectors are properly installed and maintained in homes.
- In about 40 percent of all home fires studies, smoke detectors provided the first warning of a fire. Smoke detectors cannot protect your home if a fire does break out.

Smoke detectors do not extinguish fires. A smoke detector is an early warning system that sounds an alarm that a fire is present, that something is burning.

A smoke detector can alert you to a fire in your home. This early warning increases your chances of escaping safely, but along with the early warning, it is essential that everyone have and practice a home fire escape plan.

There are two basic kinds of smoke detectors. One incorporates ionization, and the other works on the photoelectric cell principal. Either kind can sense smoke. There are arguments for and against each type of detector, but either one provides adequate warning.

The detectors that work on the ionization principal utilize a tiny bit of radioactive material to "ionize" the air in a small pocket. The radioactive material is so insignificant that it poses no danger. For example, when smoke particles, no matter how invisible, pass through the very sensitive ionized area, the electric current is interrupted and triggers the alarm.

The photoelectric detector usually contains a minute electric bulb that's barely visible at night. When smoke passes through it, the light is scattered, and a sensitive photoelectric cell triggers the alarm—a piercing wail.

### Fire! Prevention, Protection, Escape

The difference in effectiveness between each type of smoke detector is minor. The federal government's National Bureau of Standards reported that "photoelectric detectors seem to respond better to a smoldering type of fire, while ionization detectors appear to respond slightly better to flaming fire."

Either detector has a high probability of providing you with enough warning to make a safe escape. But there are questions you should ask before you buy a detector. For example, should it be battery operated or plugged into the home electric current? (Ideally, it should have at least two sources of power.)

Battery-powered detectors take less time and require fewer tools to install. The battery lasts about a year; toward the end of the life cycle, a weak battery activates the detector to emit "beeps" about every minute before the battery finally goes dead. The warning beeps, which can go on for a week, are a signal to replace the battery.

Detectors that operate off house current may require the services of an electrician to provide an electrical outlet to power the alarm. House current can stop during a fire, and the detector then becomes inoperable unless it has a backup power source.

How many detectors are required in a home or apartment? This may be a more important question than what kind of smoke detector to buy. Tests conducted by the National Bureau of Standards have shown that two detectors, on different levels of a two-story home, are twice as likely to provide an adequate amount of time for escape as one detector. Some experts believe that there should be at least one smoke detector on each level of a home, including the basement level.

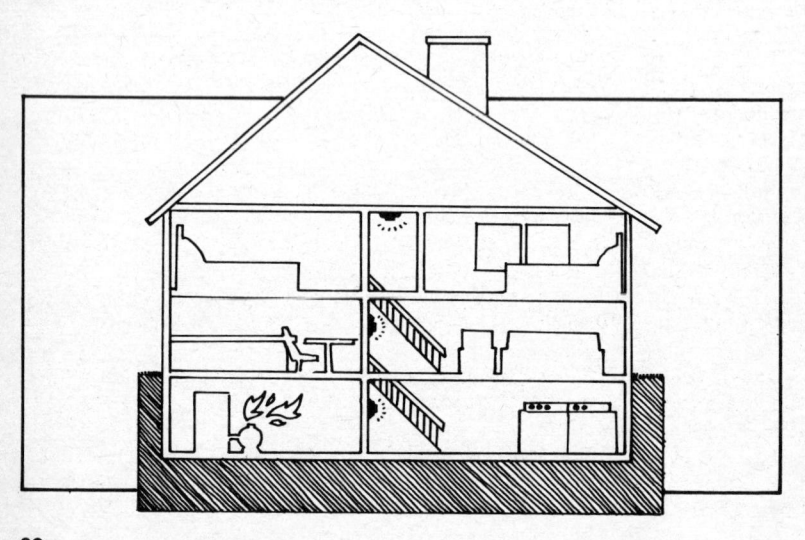

### Fighting a Home Fire

By having a minimum of two detectors, the homeowner can purchase one using the ionization principal and the other using the photoelectric cell. Two smoke detectors are far less likely, statistically, to be "on the blink" at the same time.

Smoke detectors can be purchased in hardware stores, department stores, building supply stores, and discount stores. Whatever model you buy, and wherever you buy it, it's imperative that you look for a mark or statement on the package that the unit has been tested and certified by a reliable testing laboratory.

Exactly where should smoke detectors be installed? They should be placed high on the wall, about 6 to 12 inches below the ceiling, or on the ceiling over a bedroom doorway.

Smoke detectors should never be positioned less than 12 inches from where the ceiling and wall meet, or in any corner of the room that can be a dead air space that gets little circulation. Also, the detector should not be placed in front of a wall or ceiling air duct where relatively clean air may be "washing" the detector while most of the air in the house contains smoke.

Also avoid placing detectors on a ceiling that is substantially warmer or colder than the rest of the room. In either case, an invisible "thermal barrier" near the surface can prevent the smoke from reaching the detector. This is usually a problem in mobile homes or in older, poorly insulated houses. In such cases, mounting the unit on an inside wall, 6 to 12 inches below the ceiling, will prove more reliable.

Do smoke detectors require care? They don't require much attention, but they should be tested regularly, and batteries should be replaced when necessary. Neglecting these basic requirements will render these units useless at the critical time: when a fire occurs and they are needed most.

Test them monthly by holding a burning candle 6 inches below the detector. If it's an ionization unit, let the candle flame burn; if it's a photoelectric detector, snuff the candle and let the visible smoke drift into the detector. In either case, the alarm should sound within 20 seconds.

Although some older smoke detectors have test buttons, using a candle will test all the detector's circuits. Pressing the button activates only the warning sound. Newer models have more refined functional test systems that simulate the presence of smoke.

Smoke detectors are not toys. They should be touched only for testing or battery replacement or if set off as part of a periodic family fire drill.

When buying a smoke detector, over and above making sure it carries the seal of a reputable testing laboratory, there are two important questions to ask:

1. *Can the unit be installed easily?* Do the directions consist of a step-by-step explanation with diagrams illustrating how and where to install the detector?

Fire! Prevention, Protection, Escape

2. *What maintenance is required?* Do the instructions tell how to test and how often? If the unit uses a battery or requires replacement lamps (for photoelectric model), is there a readily audible or visible signal that means it is time to replace a battery or bulb?

Special smoke detectors are available for the handicapped. One version is connected to a bed vibrator for the blind or the deaf. Another model, also for the deaf, has a very bright flashing light, similar to the flashing lights on a police car or fire engine. Your local fire department will have more information about where these special units can be purchased.

Another smoke detector is designed to protect the home when occupants are away. It is connected to a neighbor's home or to the local fire station and sounds an alert by way of a special radio transmitter.

As time goes on, smoke detectors will be available in many combinations for fire protection and safety tasks. For example, some manufacturers offer a heat-sensing device as either a standard or optional part of a detector unit or as a separate product. Most heat-sensing devices have been available for 40 to 50 years. Many use a specially formulated metal that either melts or distorts and that sets off an alarm when the change occurs. Heat detectors built into smoke detectors set off the alarm when a certain temperature is exceeded. They are valuable in an

# Fighting a Home Fire

environment that might "fool" a smoke detector, such as a kitchen, where grease particles in the air could cause the very highly sensitive ionization smoke detector to sound a false alarm.

Heat detectors can be used in areas that are too hot or too cold for some smoke detectors to function properly, such as furnace rooms, attics, and attached garages. Utilizing a heat sensor along with a smoke detector is a much more efficient warning system.

Other heat or flame detectors available on the market have both ultraviolet and infrared sensing devices. These are usually installed in industrial or business locations where highly combustible materials are stored, processed, or transported. These more expensive models will become available for the home market shortly.

## Fire Extinguishers

A basic item of fire protection equipment is the fire extinguisher, especially the smaller model that can be used in the home, garage, or home workshop or can be carried in an automobile, recreation vehicle, or boat. These extinguishers come in a basic configuration—a cylinder or tank that is usually painted red.

A fire extinguisher is generally described by the NFPA as "a storage container for an extinguishing agent such as water or chemicals, and it is designed to put out a small fire—not a big one."

Fire extinguishers are labeled according to the type of fire that might break out. Using an extinguisher with the wrong chemical against a certain kind of fire could make the fire worse. The chemical in an extinguisher classifies the type of fire against which it is to be used.

Fire extinguishers are labeled A,B,C and D. The first three letters are the most common.

- The A extinguisher contains chemicals to put out fires of paper, cloth, wood, rubber, and many types of plastics.
- The B category contains chemicals that will extinguish fire in oils, gasoline, paints, lacquers, cooking grease, solvents, and other flammable liquids.
- The C extinguisher can be directed against electrical fires in wiring, fuse boxes and other electrical units.
- The D model is for extingishing fires caused by combustible metals such as magnesium or sodium.

Different fire extinguishing chemicals have been combined for multipurpose use. For example, a home extinguisher can be labeled *ABC* or *BC*. NFPA recommends the following fire extinguishers for the home:

- The ABC multipurpose dry chemical extinguisher puts out most home fires.

Fire! Prevention, Protection, Escape

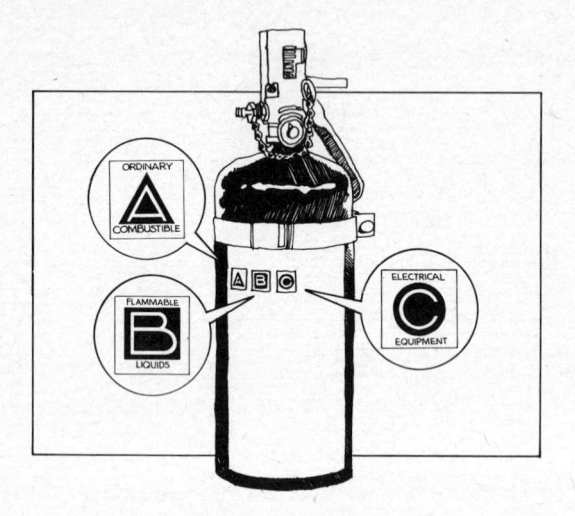

- The BC extinguisher puts out kitchen fires.
- The ABC extinguisher is for the garage or basement.

Extinguishers should be kept away from potential fire hazards. Preferably, they should be placed near an escape route in the home.

## Other Firefighters

Rope ladders can be stored in the home to help facilitate an escape in the event of fire. They can be stored in various closets or cabinets located near windows, in a flat box that fits under a bed, or in the bottom drawer of a bedroom dresser.

Along with the special ladders should be a heavy-duty flashlight and a 50-foot length of half-inch nylon rope that can support even the heaviest person. Also keep handy a small first-aid kit to treat scratches or skinned knees and arms.

Among the new fire protection and safety items that have appeared on the market in recent years is a small, portable, combination smoke/heat detector connected to a burglar alarm, which activates if it is moved. Made for the traveler, it is meant to hang from hotel room doorknobs. Another item is a ten-ounce emergency head mask that slips over the head and has a fog-proof, noncombustible transparent window and includes built-in respirator filters.

One important reminder. All kinds of fire "safety" equipment for use by travelers or in the home find their way onto the market. But these items will never replace a well-thought-out escape plan that is practiced periodically.

## ✔ Checklist for Fighting a Home Fire

☐ Does your home or apartment have a smoke detector?
☐ Have you checked your smoke detector(s) lately?
☐ Do you have a fire extinguisher in your home?
Is it readily available in case of fire?
☐ Do you have enough of the right kind of fire extinguisher?
☐ Do you know how to use your fire extinguisher?
☐ As part of your home escape plan, have you included a flashlight, and have you checked the batteries recently?
☐ Do you have a first-aid kit, rope ladder, or any other emergency items?
☐ Have you had your local fire department check the chemicals in your fire extinguisher?

Chapter 12

# Burns and Other Injuries

One of the nastiest and most painful bodily injuries is a burn. Fires, electric shock, and too much sun at the beach are just a few of the ways that burn injuries occur.

Burns can occur in the home, at work, while traveling, and during leisure time outdoors. Or occasionally, during a thunderstorm, if lightning strikes nearby.

Experienced firefighters, who are always in danger from flames and smoke and falling or crumbling debris, also suffer burn injuries.

As a result of these accumulated fire experiences, more is known today about burns and the first-aid treatment that burn injuries require.

A most common burn injury, and one that can happen anywhere, occurs when clothing catches fire. Clothing has caught fire while people have been cooking in the kitchen, at the outdoor grill, or over a campfire. In winter, burns from space heaters have contributed to the statistics.

When clothing catches fire, the natural tendency of the victim is to either beat at the flames or panic and run. Don't beat at the flames! More often than not, this will spread the flames. And don't run! Running only fans the flames; the victim can become a human torch. What should you do? Quickly shed the burning garment if it's easily removable. A man's jacket, for example, can be removed almost instantly. It takes seconds to shuck off burning trousers. On the other hand, a burning shirt cannot be unbuttoned quickly; and ripping it off takes precious time.

In fact, women's clothing poses a problem. Many dresses, such as those with buttons, can't be removed quickly enough in a fire emergency. Jackets, coats, and many robes can be shed in very short order, however.

If burning clothing cannot be removed speedily, a blanket, rug, or

## Burns and Other Injuries

coat, tightly wrapped around the victim, will smother the flames. If none of these items is readily available, don't waste time fetching them. Drop to the ground (if outdoors) or the floor (if indoors) and slowly roll so that the flames from the burning clothing are smothered by body weight. Stop, drop, and roll is easier said than done, but by pressing the burning clothing against the floor or ground surface, oxygen will be cut off and the fire's progress halted.

Dropping to the floor is a logical action if you think about it. Flames always burn in an upward direction. When you are standing, flames can burn your face and hair. This is less likely to happen if you are in a horizontal position.

It is contrary to our instincts to lie down when our clothing is on fire. The instinct is to run. Therefore, to practice fire situations that involve burning clothing can be worthwhile to the home escape plan. Burning clothing is common in fire injuries.

*Clothing can burn!* That's one of the warnings in NFPA's Learn Not to Burn! fire safety program. It recommends wearing close-fitting clothing with a tight weave, of sturdy weight and a smooth texture. Such clothing is safer than clothing of loose fit and weave, which are light in weight and fuzzy in texture and can burn quickly.

Although tighter-fitting clothing provides more protection against clothes catching fire, the NFPA warns that clothing not made of fire-retardant material, or untreated with such a retardant—such as thin cotton, silk, linen, and blends of natural and synthetic fibers—burns fast and should be avoided.

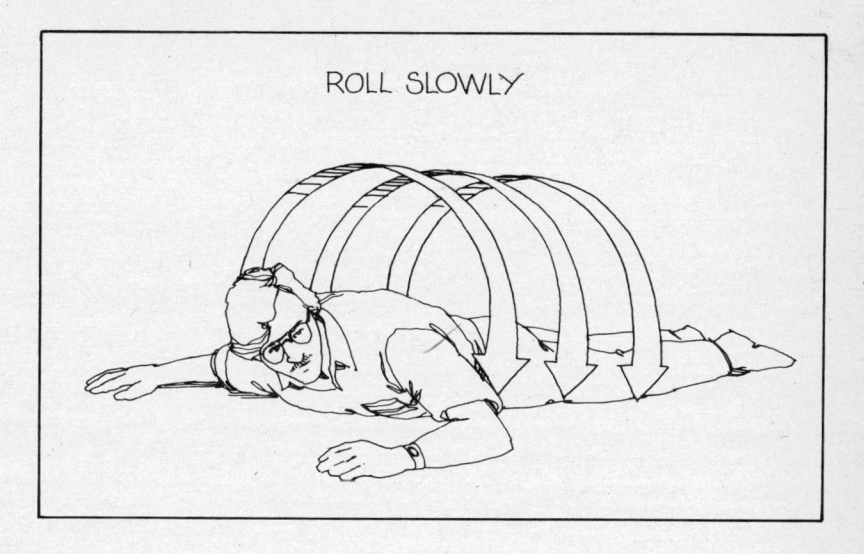

## Fire! Prevention, Protection, Escape

There are many examples of how burn injuries occur: an unsupervised toddler experiments with matches carelessly left within his reach; a child climbs on a hot stove, pulls on a pot handle, and is scalded by falling liquid; another child touches a live electrical outlet or falls against a hot heater or stove; a youngster pours gasoline on an open fire; a teenager or adult carelessly refuels a lawn mower while its motor is running; a woman prepares a hot meal on a stove while wearing a long, loose-sleeved robe; a man falls asleep while smoking in bed; an elderly person grows drowsy and drops a lighted match on her clothing.

Anyone who has suffered from a burn is aware of the sharp, throbbing pain caused by exposure to fire. "Burns and scalds are injuries caused by dry heat, fire or heated objects, steam, hot liquids, electricity, friction and chemicals," advises the National Safety Council.

Burns are classified as first degree, second degree, and third degree, according to the degree of injury to body tissue.

- *First-degree burns.* Limited to damage to the outer layer of skin, which will redden painfully.
- *Second-degree burns.* Result in damage to the outer and inner layers of the skin, which will blister painfully.
- *Third-degree burns.* Destroy the outer and inner skin layers and nerve endings. This burn category is the most dangerous, although intensive medical care and special treatment for burns have saved the lives of victims who have had up to 70 percent of their bodies injured in a fire. The initial acute pain soon disappears from the white or charred burn areas. A long healing process is required, often followed by extensive skin grafting and plastic surgery.

### Type of Burns

**First-degree burns:** caused by scalding, contact with hot objects or brief contact with fire or sunburn.

**Second-degree burns:** caused by hot liquids, gasoline flash burns, kerosene, and other chemical products; deep sunburn; all followed by

### Treatment

Relief can be administered by cold-water application or submerging burned areas in cold water. Ointments, or butter and fats, should *not* be applied. Professional medical treatment is required, especially if a doctor determines that more than 15 percent of the body is burned; same for children with 10 percent body burns. Any deep burn that blisters should be treated medically.

Relieve pain and exclude air from area by submerging burned areas in cold or ice water, apply cold packs of sterile or freshly laundered cloth wrung out in cold water, or cover burn with a wet dressing under a plastic sheet. *Never apply dry dressing* to this

## Burns and Other Injuries

early blistering of skin.

**Third-degree burns:** caused by direct contact with flame, hot objects, immersion in scalding water, or electric shock.

kind of burn as it sticks to wound. Don't break blisters or apply antiseptic preparation, ointment, grease, sprays, or home remedy. A second-degree burn requires immediate medical attention.

Requires immediate medical attention. Cover burned area with sterile dressing or freshly laundered sheet or other clean household linen. Keep burned hands higher than heart; also elevate burned feet or legs to permit liquids in burned areas to circulate. *Never* remove charred clothing or other particles from this type of burn, or permit victim to walk. Do not apply ointments, commercial burn preparations, grease, or home remedies.

If someone has received an electric burn caused by fallen live electric wires, do not touch the person's body until any live wires on the body has been removed. If the victim is unconscious, check breathing and heartbeat and immediately apply mouth-to-mouth artifical respiration and Coronary Pulmonary Resuscitation (CPR), or external heart massage. Medical assistance should be summoned immediately.

Chemical burns are as painful and dangerous as fire burns. They can be caused by household items such as drain cleaners, bleach, toilet bowl cleaners, swimming pool chemicals, auto battery acids, and any other caustic items found in the home. The first step in treating a household chemical burn is to read the instructions on the container label if the cause of the burn is known. Labels usually contain first-aid measures.

If this information is not available, immediately treat all burns to the arms, legs, and torso by washing the chemical from the skin, using lots of clean, cool water in a bathroom shower or from a hose. Wash the exposed area for at least five minutes.

*Lye or carbolic (phenol) acid burns should not be washed in water.* The approved treatment is to brush the lye from the skin and wash carbolic acid with alcohol or an alcohol derivative. Contaminated clothing should be carefully removed to avoid spreading the chemical to other parts of the body.

Chemical burns inside the mouth should be treated with lots of water. The water is not to be swallowed at first, just gargled. After at least 10 minutes, this treatment may be followed by drinking lots of milk or water.

Chemical burns to the eyes should be treated immediately by washing the eyes for at least five minutes with clean, cool water with eyelids

## Fire! Prevention, Protection, Escape

open (contact lens should be removed). Do not permit victim to rub eyes; and if particles of dry chemicals are floating on the eye, lift them off gently with sterile gauze or folded dry facial tissue—and then immediately call for medical assistance.

All chemical burns require medical treatment as soon as possible after first aid.

A seasonal nature "burn" is the severe sunburn, or "sun poisoning" whose symptoms are fever and nausea. This class of sunburn is often classified as a second-degree burn. If unusually severe, sunburns should be treated the same as bad first- and second-degree burns. Some relief from sunburn can be obtained by the application of an oil or ointment, however, to exclude air and reduce pain on the body's burned surfaces. Aspirin can reduce discomfort and fever, as can the application of soothing wet, cold packs and the drinking of lots of water.

Burns can be deceiving. They often appear less serious than they really are; and it's often difficult to distinguish between second- and third-degree burns. It helps to know about first aid for burns, but always try to obtain professional medical assistance or deliver the victim to the nearest hospital if medical aid is not available.

### ✓ Checklist for First Aid for Burns and Other Injuries

☐ Does your first-aid kit have sterile dressings and first-degree-burn medication?

☐ Have you prepared a small first-aid burn treatment instruction card for use with your first-aid kit?

☐ Is the name, address and phone number of your nearest burn center or hospital listed on your first-aid burn treatment instruction card?

☐ Have you carefully read this chapter about first aid for fire, electrical, and chemical burns?

## Chapter 13
# After the Fire

A fire breaks out in a home. It is the family of five's first experience with such a traumatic event. They will talk about it for months to come.

The fire occurs when everyone is asleep. The smoke detectors sound. Smoke fills the upstairs corridor leading to the bedrooms. Children cry out. The father awakens first and sits up in bed. He shakes his wife awake. They grab their robes, and he pulls a flashlight from the night table at his side of the bed.

He opens the bedroom door and steps into a hall filled with smoke that his flashlight beam barely penetrates. He calls the names of his three children, his voice edging on panic as he yells each name. The smoke is too much and he begins coughing and gasping. Before he can step back into his bedroom and close the door, his oldest daughter bumps into him, sobbing and frightened. As she coughs, he pulls her into the master bedroom, which is filling with smoke. He is relieved to see that she has her three-year-old sister in her arms and that her ten-year-old brother is holding on to her robe with one hand while clasping his Boy Scout flashlight in the other.

Both parents try to maintain their composure, for they know that they're in danger. Everybody is coughing and choking from the smoke, and all are trying to talk at once. The father shouts for silence as he lifts a window.

"We're going to escape just like we practiced," he tells his wife and children. His wife climbs through the window to the small outside terrace. The family is implementing a plan they developed, but never thought they would have to use. The father wonders if his wife and children will remember what they practiced on a few occasions. In the excitement, he recounts some of the details.

It is a 15-foot drop to the ground. Too high for them to jump. They had purchased a rope ladder just for this possibility. The father orders his son to get the ladder from the box in the closet.

**Fire! Prevention, Protection, Escape**

"Hurry!" he shouts, trying to keep calm. If he displays any panic, the children may panic. Panic is dangerous and contagious.

His son hands him the rope ladder, and he helps the children through the open window to the terrace and then follows them. The fresh air helps. They gulp it in as the father quickly fastens the ladder and tells his wife to climb down first. Then he helps his teenage daughter begin her descent. His wife, standing below, will assist each one to reach the final rungs to the ground. Next is his son. The youngster quickly clambers down.

The father is left with the three-year-old, who is now crying. His attempts at calming her fail. He doesn't quite know how to climb down the rope ladder with her because he will need both hands, and he's afraid to trust her to cling to him with her arms around his neck as he climbs down.

He hears the sirens growing louder. Somewhere in his subconscious he's aware that the fire department is on the way. Suddenly, the flashing lights of fire trucks appear. Floodlights are blinding him as he stands at the railing, clutching his daughter. Moments later, a helmeted figure appears at the terrace railing, shouting for him to hand over his daughter and follow the fireman down the ladder. This is a real ladder; he quickly descends to the ground. Another fireman leads him to his family. They are waiting at the escape plan's designated meeting place.

He watches firemen direct water at his home. He hears axes smash windows, and firemen shout instructions to one another. In a few minutes, it's over. He has hardly noticed the orange glow of the flames buried in the thick black smoke. He mechanically responds to the questions asked by the fire chief, the man wearing the white helmet.

It's routine, the chief says. A report has to be submitted. Regulations, you know.

The man, his wife, and their youngest child are invited into a neighbor's home. The two older children are with another neighbor. Suddenly, he feels exhausted. His robe and pajamas smell of smoke, but he's too tired to shower. He finally falls asleep. It's four a.m.

About two hours later, he is awake. The sun is up, and it's going to be a glorious day—for others, not for him. He throws on his robe and steps out into the sunlight so that he can see his house across the street. He winces at the sight of shattered windows and charred siding. He walks across the front lawn, still soggy from the water directed at the house by the firefighters.

The siding is not charred, those are smoke marks. But the windows. They need more than panes of glass. Inside is the smell of burned wood and the pungent odor of burned upholstery. He shakes his head. He has survived a fire in his home. But what does he do now?

Questions run through his mind. There's the insurance policy, the insurance company has to be called. He wants the adjuster to examine the damage as soon as possible. Try as he might, he can't remember the terms of the fire insurance policy. And he is going to have to wait

# After the Fire

until nine o'clock when the bank opens. The policy is with other valuable documents in a safety deposit box.

The inside of the house looks like it went through a war. The dining room is a mess. The fire reached the study. The staircase is badly burned, and he wonders if the charred stairs will hold his weight. His clothes are in his bedroom. And what about valuables? He decides not to try to climb the stairs.

Suddenly he hears a sob behind him. It's his wife. She stands in the front doorway, wringing her hands. He steps back to her side and holds her in his arms. It's going to be a long and busy day.

After telephoning his insurance broker, he visits the bank for the documents that he'll need. When he returns home, he finds a fire marshal talking with his wife. The fire marshal tells them about local disaster relief services, such as the Red Cross and Salvation Army, which assist insured as well as uninsured fire victims. Although they don't think that they will need most of the available public services, the fire marshal advises that these agencies can help with food, replacement eyeglasses, or medicines destroyed in the fire; and that such emergency relief is given free of charge and regardless of income.

The fire marshal also warns that care should be taken. For example, the fire can rekindle from hidden, smoldering debris. Stay clear of all wiring that may have been water damaged, he cautions. And look out for structural damage. The roof and floors may have been weakened.

The fire marshal also advises that the utilities will be shut off until it's determined that they are safe to use. And he warns them not to try to connect utilities without professional help.

After the fire marshal departs, they try to read the fine print on the insurance policy. The insurance company will provide the family with temporary lodging. Perhaps there's an option to remain in the house, but it will have to await the decision of the town's building inspector, who also must be contacted. There's so much to do.

During the first day after a fire, there are a number of steps to take:

- If a home has been damaged in a fire and cannot be used, advise the local police department to keep a protective eye on the property.
- See if your insurance company provides an immediate cash advance to the insured as part of the final settlement.
- Keep receipts of out-of-pocket expenses related to a fire loss for reimbursement by the insurance comany.
- When sifting through the remains of a fire-damaged home, take care to look for valuable papers and one or more forms of identification.
- Look for other valuables, such as credit cards, checkbooks, insurance policies, savings account bankbooks, money, and jewelry.
- If there is a wall or floor safe, do not attempt to open it for at

least ten hours. A safe that has been through a fire may hold intense heat for several hours. If the door is opened before the metal has cooled down, the air entering, combined with the high temperature within, may cause the paper contents of the safe to burst into flames.
- Collect eyeglasses, hearing aids, prosthetic devices, and other personal aids. Check out the condition of all prescription medication, such as insulin, blood-pressure-regulating drugs, and drugs used for cardiac and other physical ailments.
- Protect property from further damage by having immediate and reasonable repairs made, such as covering holes in the roof or walls. Take reasonable precautions against further damage and loss, such as draining water lines in winter if the house is to go unheated for a period of time. Temporarily repair any additional damage from the elements, such as frozen and burst pipes and leaks of rainwater. Additional losses are often unacceptable to insurance companies, who refuse to pay for losses for which no "reasonable" care was taken.
- Inform the mortgage company. The homeowner must tell the mortgage company of fire damage, and must keep the company informed of plans to restore the property. The mortgage holder "owns" a portion of the dwelling and is interested in knowing that its investment is handled properly. Mortgage holders also will provide forms to fill out; some may insist on inspecting the premises. The names of both the mortgage holder and the homeowner will be on the face of the check paid by the insurance company to cover the cost of repairs. It's to the advantage of fire victims who own homes to work together with their insurance company and mortgage holder.
- Alert the landlord, if you live in an apartment. Apartment dwellers who have been through a destructive fire should be aware that it's the landlord's responsibility to prevent further loss at the fire site. But it's up to the resident of the apartment to see to the security of personal belongings, such as clothing, furniture, and valuables, by moving them to another location in the building or to the home of a friend or to storage. Many apartment residents also maintain fire insurance policies; in the event of a fire, their insurance companies should be informed.
- Do not discard damaged clothing, furniture, or other personal belongings until a thorough inventory of the destruction is completed. All damages are taken into consideration in the development of insurance claims.
- Consult the insurance company or adjuster before seeking repair and inventorying estimates.
- Take an inventory. Insurance companies often request property

## After the Fire

insurance clients to inventory their personal belongings. Almost no one does. But those who do receive faster service from their insurance companies. One way of supporting personal claims is to make a prefire inventory, which should include photos taken at home in happier days that show furnishings and other belongings. This kind of inventory can help immeasureably in the claim estimating procedure.

- Don't put off the necessary work. The aftermath of a home fire can be traumatic, and the inventory process can be difficult and emotionally upsetting. But time must be given to this chore.
- Timely action can reduce additional loss. A small tape recorder may be helpful in creating a list of destroyed or damaged items, along with observations and comments, especially recollections of original costs. Insurance agents or brokers can provide the inventory forms.
- Look for receipts. Receipts are helpful in establishing original value, but insurers understand that in most cases policy holders cannot provide this type of documentation for damaged or destroyed items. Receipts may have been destroyed in the fire. Nonetheless, both insurer and insured must agree on the value of loss. If no agreement is possible, provision is made in the policy for the appraisal of loss and the arbitration of differences.
- Consider your options. Some insurance policies give the *insurer* the option to repair or replace the damaged or destroyed item, whichever costs less. Other policies give this option to the *insured*. In either case, when the dollar value of the property has been transferred to the owner, the property belongs to the insurer and may be disposed of for its salvage value.
- Remember that once an insurance check for loss is cashed, that portion of the settlement has been accepted by the insured party.
- Consider your tax position. Fire losses are deductible from state or federal income taxes. Maintain detailed records, and keep receipts paid for the repair and replacement of damaged or destroyed property, and the cost of living expenses in another location during the period of residence. If the loss in one year is larger than income, there may be a hefty income tax refund due. For additional information, contact your local Internal Revenue Service office and request a copy of Publication 547, *Tax Information on Disasters, Casualty Losses and Thefts*.
- Ask the local fire department for further information and aid in the recovery process. The local fire department should have copies on hand of the Federal Emergency Management Agency pamphlet *After the Fire: Returning to Normal*.

### ✔ Checklist for Information Required After a Fire

☐ Do you know where your home insurance policy is located? Are you aware of the terms?

☐ Have you made an inventory of your personal belongings and their estimated value?

☐ Do you have purchase receipts in a safe place for valuable items of jewelry, furs, paintings, antique furniture, and the like?

# Index

**A**
Aftermath of fire, 73-76
Animals, and fires, 15-16
Appliances, electrical, safety with, 9, 11
Attics, 5
Automobile exhaust pipes, 52

**B**
Babysitters, instructing, 29-31
Baker, Howard, xiii
Baking soda, for extinguishing fires
  barbecues, 46
  in the kitchen, 9, 10
Barbecues, 46-47
Bedroom, "fuel" in, 3
Boating safety, 50-52
  chores checklist, 50
  fire extinguisher for, 52
  fire safety checklist, 56
  fueling, 51
Burglar alarm and smoke/heat detector, 64
Burning,
  leaves, 45-46
  vacuum cleaner bags filled with dust, 41
Burns
  chemical, 69-70
  classification of, 68-69
  from clothing fire, 66
  electric, 69
  treatment for, 68-70
Bush, George, xiii
Butane, 51

**C**
Campfires, 53-54
Camping safety, 52-54
  and fire safety checklist, 56
Candles, 5, 48
Carbolic (phenol) acid burn, 69
Carter, Rosalyn, xiii
Cellars, 5
Channing, Carol, xiii
Charcoal starter fluid, 46
Chemical burns, 69-70
Chicago fire, x
Children
  causing fires, 28, 29
  fire escape plan checklist for, 33
  instructing babysitters of, 29-31
  and matches, 4, 28, 29-30
  teaching fire prevention to, 28-29
Chimney fire, 43
Christmas hazards, 43-44
  tree lights, 40
  trees, 43-44
Chronology of U.S. fires, x
Cleaning rags, storing, 5
Clothing, on fire, 66-67
  burns from, 66
  removing, 66-67
  smothering flames on, 66-67
Cocoanut Grove niteclub (Boston) fire, x
Cold-weather hazards, 39-44
Cole, Natalie, 34
Collectibles stored away, 5-6
Cooking grease fire, 13, 63
Cooking oil fire, 8-9, 63

# Fire! Prevention, Protection, Escape

Creosote, 39, 43
Curtains, kitchen, 9

## D

Davidson, John, xiii
DeBakey, Dr. Michael, xiii
Doors
  blowing open, 23
  escaping fire through, 23, 24
  pressure caused by superheated air, 23, 24
  as safety barrier, 23

## E

Elderly
  and causes of fires, 31-32
  fire escape plan checklist for, 33
  fire safety for, 32
Electric burns, 69
Electric heaters, 40, 42
Electric light bulbs, bare, 5
Electric saws, 46
Electrical appliances, safety with, 9, 11
Electrical cords, 11
Electrical fires, 10, 63
Electrical storms, 54
Electrical system, 10-11
Elevators, 34, 38
Equipment, safety with, 46
Escaping fire, 71-72
  checklist, 27
  children, 33
  elderly, 33
  and home escape plan, 17-22
  from hotel room, 34, 36-38
  through doors, 23, 24
  through smoke, 26
  through windows, 25-26
Exhaust system, in the kitchen, 12-13
Extinguishers, fire, *see* Fire extinguishers

## F

Federal Emergency Management Agency, 59
Fighting a home fire
  checklist for, 65
  determining whether or not to fight, 57-58
Fire extinguishers, 63-64
  for home fires, 63-64
  placement of, 63
  types of, 63
Fire hazards
  animals as, 15-16
  and children, 28, 29
  and the elderly, 31-32
  in the home, 4-14, 44-45
  recreational, 50-55
  seasonal, 39-49
Fire-starter fluids, 46
Firecrackers, 47-48
Fireplace damper, 43
Fireplaces, 42-43
Fires
  aftermath of, 73-76
  checklist for information required after, 76
  escaping, *see* Escaping fire
  ingredients for, 3
  number of each year, xi
  reported, x-xi
  temperature of, 4
  unreported, xi
Fireworks, 47-48
First-aid for burns, 68-70
First-aid kit, 36, 64
First-degree burns, 68
Flame detectors, 63
Flammable liquids
  fire extinguisher for, 63
  safety with, 13-14
Flashlights, 35, 64
Floor waxes, storing, 5
Freak incidents, causing fires, 16
Furniture polish, storing, 5

## G

Galley stoves, 1, 51
Garages, 5
Gasoline
  vapor fuse, 13
  storing or carrying, 52
Grease fire, 8-9, 63

# Index

in exhaust system over stove, 13
Ground metal, 45

## H
Halloween hazards, 48
Heat detectors, 63
Heat-sensing device, 62-63
High-rise building fire safety checklist, 38
History of fires, x
Holiday hazards
  Christmas, 43-44
  Halloween, 48
Home
  aftermath of fire in the, 73-76
  checklist for eliminating dangerous habits, 7, 14
  checklist for fighting fire, 65
  electrical system in the, 10-11
  fire extinguishers for the, 63-64
  fire hazards in the, 4-14, 44-45
  fire safety precautions before leaving for vacation, 52
  "fuel" for fire in the, 3
  heat detector for the, 63
  ingredients for fire in the, 3
  and pets as fire hazard, 15-16
  safety equipment for the, 64
  smoke detectors for the, 59-62
  spontaneous combustion fires in the, 4-6
  storing collectibles in the, 5-6
  storing household cleaners in the, 4-5
  workshop safety, 44-45
  *see also* Kitchen
Home escape plan, 17-22
  checklist, 22
  importance of, 17-18
  individual responsibilities in, 21
  preparation of, 18
  rehearsal, 21
Home workshop safety, 44-45
Hotels
  escaping fire in, 34, 36-38
  fire safety checklist, 38
  fires, 34
  and travelers' checklist of items to carry, 35-36
  and travelers' room checklist, 36
Household cleaners

causing spontaneous combustion, 4-5
storing 4-5

## I
International Association of Fire Chiefs, xiii
International Association of Fire Fighters, xiii

## J
Jack 'o lanterns, 48
Jamestown (Virginia) settlement fire, x

## K
Kelly, Gene, xiii
Kerosene, and vapor fuse, 13
Kerosene heaters, 39, 42
Kitchen, 8-10
  baking soda for cooking oil fires, 9, 10
  curtains, 9
  electrical appliances in the, 9
  heat-sensing device for, 62-63
  oven fires, 9
  safety, 9-10
  stove, 8-9
  ventilation system fire, 12-13

## L
Ladders, rope, 64, 71
Lane, Dick, 34
Las Vegas hotel fire, 34
Lawn mowers, 46
Leaves, burning, 45-46
Light bulbs, bare, 5
Lightning, 54
Liquefied petroleum gas (LPG)
  barbecue, safety with, 47
  for boats, 51
  fire, extinguishing, 52
Living room, "fuel" in, 3
Lye acid burn, 69

## M
Manufactured homes, 55-56
Marine fire safety, 50-52
Matches

# Fire! Prevention, Protection, Escape

and children, 4, 28, 29-30
extinguishing, 4
safety with, 4
Metal dust, 45
MGM Hotel (Las Vegas) fire, x
Minnesota fire, x
Mobile homes, 55-56
Mops, storing, 5
Motels
  and fire safety, 38
  fires, 34

## N

Naptha, and vapor fuse, 13
National Bureau of Standards, 60
National Fire Protection Association (NFPA), xiii, 59, 63
  *Fire Protection Handbook*, 10
New Orleans fire, x
NFPA, *see* National Fire Protection Association
Nixon, Richard, xiii
Nylon rope, 64

## O

Oil, cooking, 8-9
O'Leary, Mrs., x
Oven fires, 9

## P

Paint removers, storing, 5
Paints, storing, 4, 5
Panic, 72
Pets, and fires, 15-16
Polishing cloths, storing, 5
Portland, Maine fire, x
Power tools, 46
Propane, 51
Prowse, Juliet, 34
Pumpkins, Halloween, 48

## R

Recreation
  boating, 50-52, 56
  camping, 52-54, 56
Recreational vehicles, 54-55, 56
Rope, nylon, 64
Rope ladders, 64, 71

## S

Sawdust, 45
Scalding, 68
Seasonal fires, 39-49
  Christmas, 43-44
  Halloween, 48
  fire safety checklist for, 49
  summer, 46-48
  winter, 39-44
Second-degree burns, 68-69
Senior citizens, and fire, 31-33
Sitters, instructing, 29-31
Skyrockets, 48
Smoke, escaping fire through, 26
Smoke detectors, 59-62
  importance of, 59
  installing, 61
  kinds of, 59-60
  portable, 35-36, 64
  purchasing, 61-62
  requirements in the home, 18, 60-61
  testing, 61
Smoke/heat detector and burglar alarm, 64
Smoking in bed, 31
Sparklers, 48
Spontaneous combustion, in the home
  from household cleaners improperly stored, 4-5
  from items stored in enclosed places for long periods of time, 5-6
Stairways, and fire, 23
Statistics on fires, xi, xii
Stauffer Inn (White Plains) fire, x
Storage areas, care of, 5-6
Storing household cleaning products, 4-5
Stove
  kitchen, 8-9
  galley, 51
Summer hazards, 46-48, 50-54
Sun poisoning, 70
Sunburn, 68-69, 70
Superheated air, and effect on doors, 23, 24

## T

Temperature, of fire, 4
Tent fires, 52-53
Texas City explosion and fire, x
Third-degree burns, 68, 69

## Index

Thunderstorms, 54
Tools, safety with, 46
Travelers
 fire safety checklist, 38
 hotel checklist, items to carry, 35-36
 hotel room checklist, 36
 in hotel room during fire, 36-38
Triangle Shirtwaist Company (NYC) fire, x

### U
Underwired homes, 10
U.S. Fire Administration (USFA), 59
USFA, *see* U.S. Fire Administration

### V
Vacation, home fire safety precautions before leaving for, 52

Vacuum cleaner bags, full, burning, 41
"Vapor fuse" fire, 13-14
Vapors, 45
Varnish, storing, 5
Volatile liquids, safety with, 13-14

### W
Warm-weather hazards, 46-48, 50-54
Waxes, floor, storing, 5
Williams, Andy, 34
Windows, escaping fire through, 25-26
Winfield, Dave, xiii
Winter hazards, 39-44
Wiring, electrical, 10-11
Wood-burning stoves, 39, 40-41
Wood shavings, 45
Workplace fire, x

## WORLD ALMANAC PUBLICATIONS

200 Park Avenue
Department B
New York, New York 10166

Please send me, postpaid, the books checked below:

☐ THE WORLD ALMANAC AND BOOK OF FACTS 1985 . . . . . . . . $4.95
☐ THE WORLD ALMANAC EXECUTIVE APPOINTMENT BOOK 1985 . . $17.95
☐ THE WORLD ALMANAC BOOK OF WORLD WAR II . . . . . . . . $10.95
☐ THE WORLD ALMANAC DICTIONARY OF DATES . . . . . . . . . $8.95
☐ THE LAST TIME WHEN . . . . . . . . . . . . . . . . . . . . $8.95
☐ WORLD DATA . . . . . . . . . . . . . . . . . . . . . . . . $9.95
☐ THE CIVIL WAR ALMANAC . . . . . . . . . . . . . . . . . . $10.95
☐ THE OMNI FUTURE ALMANAC . . . . . . . . . . . . . . . . . $8.95
☐ THE LANGUAGE OF SPORT . . . . . . . . . . . . . . . . . . $7.95
☐ THE COOK'S ALMANAC . . . . . . . . . . . . . . . . . . . . $8.95
☐ THE GREAT JOHN L . . . . . . . . . . . . . . . . . . . . . $3.95
☐ MOONLIGHTING WITH YOUR PERSONAL COMPUTER . . . . . . . . $7.95
☐ SOCIAL SECURITY & YOU: WHAT'S NEW WHAT'S TRUE . . . . . $2.95
☐ KNOW YOUR OWN PSI-Q . . . . . . . . . . . . . . . . . . . $8.95
☐ HOW TO TALK MONEY . . . . . . . . . . . . . . . . . . . . $7.95
☐ THE DIETER'S ALMANAC . . . . . . . . . . . . . . . . . . . $7.95
☐ THE TWENTIETH CENTURY ALMANAC (hardcover) . . . . . . . $24.95
☐ THE COMPLETE DR. SALK . . . . . . . . . . . . . . . . . . $8.95
☐ THE WORLD ALMANAC REAL PUZZLE BOOK . . . . . . . . . . . $2.95
☐ ABRACADABRA: MAGIC AND OTHER TRICKS (juvenile) . . . . . $5.95
☐ CUT YOUR OWN TAXES & SAVE 1985 . . . . . . . . . . . . . $2.95
☐ MIDDLE EAST REVIEW 1984 . . . . . . . . . . . . . . . . . $24.95
☐ ASIA & PACIFIC 1984 . . . . . . . . . . . . . . . . . . . $24.95
☐ LATIN AMERICA & CARIBBEAN 1984 . . . . . . . . . . . . . $24.95
☐ AFRICA GUIDE 1984 . . . . . . . . . . . . . . . . . . . . $24.95

(Add $1 postage and handling for the first book, plus 50 cents for each additional book ordered.)

Enclosed is my check or money order for $_____

NAME_____

ADDRESS_____

CITY_____STATE_____ZIP_____